T0207184

Universitext

Universitext

Universitext is a series of textbooks that presents material from a wide variety of mathematical disciplines at master's level and beyond. The books, often well class-tested by their author, may have an informal, personal even experimental approach to their subject matter. Some of the most successful and established books in the series have evolved through several editions, always following the evolution of teaching curricula, to very polished texts.

Thus as research topics trickle down into graduate-level teaching, first textbooks written for new, cutting-edge courses may make their way into *Universitext*.

More information about this series at http://www.springer.com/series/223

Rida Laraki · Jérôme Renault ·
Sylvain Sorin

Mathematical Foundations of Game Theory

 Springer

Rida Laraki
CNRS
Lamsade Université de
Paris Dauphine-PSL
Paris, France

Department of Computer Science
University of Liverpool
Liverpool, UK

Sylvain Sorin
Institut de Mathématiques de Jussieu-PRG
Sorbonne Université
Paris, France

Jérôme Renault
Toulouse School of Economics
Université Toulouse 1 Capitole
Toulouse, France

ISSN 0172-5939   ISSN 2191-6675   (electronic)
Universitext
ISBN 978-3-030-26645-5   ISBN 978-3-030-26646-2   (eBook)
https://doi.org/10.1007/978-3-030-26646-2

Mathematics Subject Classification (2010): 91-01, 91A05, 91A06, 91A10, 91A15, 91A18, 91A20, 91A22, 91A25, 91A26, 91A50

Original French edition published by Les Éditions de l'École Polytechnique, Palaiseau, France

This Springer imprint is published by the registered company Springer Nature Switzerland AG
The registered company address is: Gewerbestrasse 11, 6330 Cham, Switzerland

Preface

Overview

Game theory is a field that studies interactive decisions. On the one hand, it builds models that represent situations where several entities or players make choices, and where this collection of individual behaviors induces an outcome that affects all of them. On the other hand, since the agents may evaluate these outcomes differently, game theory is also the study of rational behavior within this framework. The most common paradigm is defined by the range between conflict and cooperation.

The first mathematical analysis in this direction goes back to the eighteenth century and concerns the probability of winning a card game as the play unfolds (Montmort [147]). However, the foundations of game theory appeared at the beginning of the twentieth century with the works of Zermelo [231], who analyzed finite games with perfect information; Borel [31], who investigated strategies and information; and von Neumann [224], who established the minmax theorem. A major advance was achieved with the publication of the book "Theory of Games and Economic Behavior" (1944), co-authored by the mathematician von Neumann and the economist Morgenstern, which gives the first formal descriptions of several classes of games and provides the main principles of the theory. A further step was reached with the work of Nash around 1950 concerning the definition and existence of "strategic equilibrium", which among other things generalizes Cournot's work [38].

Since then, the theory of games has developed in several directions, including games with infinite action spaces, a continuum of players, dynamic games, differential games, games with incomplete information and stochastic games,.... Numerous conceptual advances have been made: refinement/selection of equilibria, new notions of solutions such as correlated equilibrium, uniform properties in repeated interactions, learning procedures,.... A wide range of mathematical tools has been employed, such as convex and functional analysis, optimization, stochastic processes, real algebraic geometry, algebraic topology, fixed-point theory, dynamical systems,.... Game theory has been successfully applied to numerous other disciplines, principally economics, biology, and computer science, but also operations research, political science, as well as other areas in mathematics (logic, descriptive set theory, combinatorial games,...).

Game theory has greatly influenced economics and several game theorists have received the "Nobel prize in economics": Harsanyi, Nash and Selten in 1994; Aumann and Schelling in 2005; Hurwicz, Maskin and Myerson in 2007; and Roth and Shapley in 2012. One could also mention Tirole in 2014 and Hart and Holmstrom in 2016, who all used and developed game-theoretical tools. Actually, most topics in industrial organization or in auction theory contain a game-theoretical aspect: the aim being to analyze the strategic behavior of rational agents acting in an environment of similar participants, taking into account idiosyncratic aspects of information.

The link with biology, initiated by Maynard Smith [130], has produced a lot of new advances, in terms of concepts, like "ESS" (evolutionary stable strategies), and of mathematical tools, in particular the use of dynamical systems. A fascinating challenge for research is to understand the similarity between outcomes induced by myopic mechanisms (like those that are fitness-based, which do not make any assumptions about the environment) and those obtained through rational deduction from clever and sophisticated participants.

A recent and very active field of investigation is the interaction of game theory with computer science. This domain has even produced the new terminology "algorithmic game theory" and was recently recognized by the Nevanlinna Prize (ICM 2018), awarded to Daskalakis. Interesting questions arise in relation to the complexity of producing efficient frameworks for strategic interactions (algorithmic mechanism design), the complexity of finding an equilibrium, or the complexity of the equilibrium itself (as implemented by an automaton).

Since the work of von Neumann and Morgenstern, games have been classified into two categories: strategic (or non-cooperative) versus coalitional (or coopera-tive). In this book, we will mainly consider the former, which offers a unified presentation involving a large variety of mathematical tools. The simplest way of representing the model is to explicitly describe the "rules of the game", as in poker or chess: which player plays and when, which information he receives during the play, what set of choices he has at his disposal and what the consequences of a play are until a terminal outcome. This corresponds to the so-called "extensive form" of the game, which allows us to introduce the notion of a strategy for each player, as a specification of her behavior at each instance where she may have to make a choice. A strategy is thus a computer program or an algorithm, i.e., a set of instructions which allows one to play the game in all possible scenarios. Clearly, by construction, if each player chooses such a strategy, a play of the game is generated, and as a consequence, an outcome is produced.

An abstract way of looking at the previous situation consists in introducing the set of strategies of each player and a mapping that associates to every profile of strategies (one for every player) the outcome corresponding to the induced play. This outcome finally determines for each player a payoff corresponding to her evaluation. This corresponds to the "normal form" of the game. Both the extensive and normal form model the strategic behavior of the agents. The extensive form is simpler to work with and usually leads to computations with lower complexity. The advantage of the normal form is that it presents a simple and universal mathematical formulation: a game with n players is essentially described by a map from a product of n spaces to $\mathbb{R}^n$.

Once an interactive decision situation has been expressed as a game (identification of the players, specification of the strategies, determination of the outcomes, and evaluation of the payoffs), one has to study what constitutes "rational" behavior in such a framework. This analysis has two components:

- the first one is conceptual and defines mathematical objects or solution concepts that take into account the fact that the players make choices, have preferences, and have some knowledge or observation about the game and other players preferences or choices. This aims to extend to interactive situations the behavior of a single agent maximizing her utility. Basic examples are the notion of value for two-person zero-sum games and of equilibrium for n-person games, but also analyses of belief formation and learning procedures.
- the second is to study properties of these objects in terms of existence and dependence of the parameters, algorithms to compute them or learning procedures that converge to them, alternative representations, extensions to more general frameworks, etc. This will be the main focus of this volume. The principal results concerning value and equilibria will be obtained by analyzing the normal form of the game. For multi-stage games and refinement of equilibria, the extensive form will be useful.

To conclude, the purpose of this book is to present a short but comprehensive exposition of the main mathematical aspects of strategic games. This includes a presentation of the general framework and of the basic concepts as well as complete proofs of the principal results.

This will give the newcomer a global overview of the fundamental achievements of the field. It will also provide the reader with the knowledge and tools needed to develop her research: either toward application to economics, biology, computer science,... or to more advanced or specialized topics in game theory such as dynamic games, mean field games, strategic learning, and its connection to online algorithms.

Summary of the Book

Chapter 1 is devoted to a general introduction to strategic interactions. A collection of examples (matching, bargaining, congestion, auctions, vote, evolution, repetition, etc.) shows the variety of situations involved and the kind of problems that occur. The formal definition of a game, notations, and basic concepts are introduced.

We then consider zero-sum games in normal form (Chaps. 2 and 3). They model pure competition between two players having opposing interests as follows: given a real-valued payoff function g defined on a product of strategy sets $S \times T$, player 1 controls the first variable and wants to maximize g whereas simultaneously player 2 controls the second variable and wants to minimize g. The corresponding solution concepts are the notions of value ($v = sup_{s \in S} inf_{t \in T} g(s, t) = inf_{t \in T} sup_{s \in S} g(s, t)$) and of optimal strategies.

In Chap. 2, we study finite zero-sum games, where the sets of strategies are finite. We prove the minmax theorem of von Neumann, which states that if a game is played with mixed strategies (probability distribution on strategies), then the value exists, as well as an optimal mixed strategy for each player. We then consider extensions such as Loomis' theorem and Ville's theorem. Finally, we introduce and study the convergence of the process "Fictitious Play", where the initial game is repeated and each player plays at every period a best response to the average of the strategies played by his opponent in the past.

Chapter 3 considers general zero-sum games and proves various minmax theorems. We start with Sion's theorem with geometrical and topological assumptions on the game: convex compact strategy sets and payoff functions quasi-concave upper semi-continuous in the first variable and quasi-convex lower semi-continuous in the second variable. Then we prove the standard minmax theorem in mixed strategies for measurable bounded payoff functions, extending von Neumann's theorem, under topological assumptions: compact Hausdorff strategy sets and payoff functions u.s.c. in the first variable and l.s.c. in the second variable. We conclude the chapter with the introduction of the value operator and the notion of a derived game, which play an important role in the operator approach to repeated games.

Chapter 4 deals with N-player games (general-sum games). Each player $i = 1, \ldots, N$ has his own payoff function g^i defined on a product of strategy sets $S = S^1 \times \cdots \times S^n$, and controls the i-th variable in S^i. The standard notions of best response, weak and strict dominations are introduced. We then define the concept of rationalizable strategies and prove an existence theorem (Bernheim and Pearce). Next we present the central notion of Nash (ε-)equilibria. The existence of a mixed strategy Nash equilibrium in finite games is then proved, using Brouwer's fixed-point theorem. Section 4.7 deals with generalizations to continuous games. The existence of an equilibrium is proved under topological and geometrical assumptions, for compact, continuous, and quasi-concave games, and the existence of a mixed equilibrium is shown under the topological conditions. Then, the characterization and uniqueness of a Nash equilibrium are presented for smooth games where the strategy sets are convex subsets of Hilbert spaces. Section 4.8 deals with Nash and approximate equilibria in discontinuous games and notably studies Reny's better-reply security. In Sect. 4.9, we explain that the set of mixed Nash equilibria of a finite game is semi-algebraic and define the Nash components of a game. Lastly, Sect. 4.10 discusses several notions (feasible payoffs, punishment level, threat point, focal point, Nash versus prudent behavior, common knowledge of the game) and Sect. 4.11 proves the standard fixed-point theorems of Brouwer (via Sperner's Lemma) and Kakutani (for a multi-valued mapping from a convex compact subset of a normed vector space to itself).

Chapter 5 describes concave games, potential games, population games, and Nash/Wardrop equilibria, and introduces the characterization of equilibria via variational inequalities. We then study the equilibrium manifold, where the finite strategy sets are fixed and the payoff functions vary. We prove the structure theorem of Kohlberg and Mertens, then show that every game has an essential component of Nash equilibria, and that for generic payoffs, the set of mixed equilibria is finite and its cardinality is odd. We also introduce Nash vector fields and game dynamics, such as the replicator dynamics, the Smith dynamics, the best response dynamics, and the Brown–von Neumann–Nash dynamics. We conclude Chap. 5 with an analysis of evolutionary stable strategies and their connection with replicator dynamics.

Chapter 6 deals with games in extensive form. Here an explicit evolution of the interaction is given, describing precisely when each player plays, what are the feasible actions, and what information is available to each player when he makes a decision. We start with games with perfect information (such as chess) and prove Zermelo's theorem for finite games. We then consider infinite games à la Gale–Stewart: we show that open games are determined and that under the axiom of choice, there exists an undetermined game. Next we introduce games with imperfect information and prove Kuhn's theorem, which states that mixed and behavioral strategies are equivalent in games with perfect recall. We present the standard characterization of Nash equilibria in behavioral strategies and introduce the basic refinements of Nash equilibria in extensive-form games: subgame-perfection, Bayesian perfect and sequential equilibria, which impose rational behaviors not only on the equilibrium path but also off-path. We prove the existence of sequential equilibrium (Kreps and Wilson). For normal form games as in Chap. 4, we introduce the standard refinements of Nash equilibrium: perfect equilibrium (Selten) and proper equilibrium (Myerson). We prove that a proper equilibrium of a normal form game G induces a sequential equilibrium in every extensive-form game with perfect recall having G as normal form. Finally, we discuss forward induction and stability (Kohlberg and Mertens).

Chapter 7 starts with the concept of correlated equilibrium due to Aumann, an extension of Nash equilibrium with interesting properties from a strategic, geometric, and dynamics viewpoint. In a learning framework, we then introduce no-regret procedures and calibrated strategies, and prove existence results. Next we show that in a repeated game: (1) if a player i follows a strategy with no external regret, the empirical distribution of moves converges a.s. to the corresponding Hannan set for i, and (2) if each player follows a procedure with no internal regret, convergence to the set of correlated equilibrium distributions occurs. We conclude Chap. 7 with games with incomplete information (Bayesian games), where the players have different information on the game they have to play, and we introduce the corresponding extension of equilibria.

Finally, Chap. 8 deals with repeated games. It first considers the simplest model (with observed moves) and introduces: finitely repeated games where the payoff is the average of the stage payoffs over finitely many stages, discounted games where the payoff is the (infinite) discounted sum of the stage payoffs, and uniform games where players consider the payoff in any long enough game (or in any discounted game with low enough discount factor). We present the Folk theorem, which states that the equilibrium payoffs of a uniform game are precisely the payoffs which are feasible (achievable) and individually rational (above the punishment level for each player). The message is simple: if players are patient enough, any reasonable payoff can be achieved at equilibrium. Folk theorems for discounted games and finitely repeated games are also exposed and proved. The last section presents extensions of the model in three directions: repeated games with signals (imperfect observation of moves), stochastic games where the "Big Match" is studied, and repeated games with incomplete information where we prove the cav$u$ theorem of Aumann and Maschler.

At the end of each chapter, we present exercises which often contain complementary results. Hints to the solutions may be found in the last chapter. The topics include stable matchings, Vickrey auctions, Blackwell approachability, Farkas' lemma, duality in linear programming, duels, Cournot competition, supermodular games and Tarski's fixed-point theorem, convex games, potential games, dissipative games, fictitious play and the Shapley triangle, the game of Chomp, a poker game, bargaining, a double auction, the possibly negative value of information, strategic transmission of information, correlated equilibrium distribution via minmax, comparison between correlated and Nash equilibria, the prisoner's dilemma with a blind player, the battle of the sexes in the dark, jointly rational payoffs, subgame-perfect equilibrium payoffs for discounted games and quitting games.

Prerequisites

This book has been written mainly for graduate students and academics. Growing from lecture notes for masters courses given in the mathematics departments of several universities (including UPMC-Paris 6 and Ecole Polytechnique), it gives a formal presentation of the mathematical foundations of Game Theory and is largely self-contained, as many proofs use tools and concepts defined in the text. Knowledge of mathematics at the undergraduate level such as basic notions of analysis (continuity, compactness), of linear algebra and geometry (matrices, scalar product, convexity), and discrete probabilities (expectation, independence) is required. Combined with an interest in strategic thinking and a taste for formalism, this will be enough to read, understand, and (we hope) enjoy most of this book.

Some proofs contain relatively more advanced material. For instance, differential inclusions are used to illustrate the convergence of the "fictitious play" process in Chap. 2. We apply the Hahn–Banach separation theorem for convex sets in Chap. 3. Borel probabilities on a compact Hausdorff space, and the weak* topology over the set of such probabilities, are used in Sect. 3.3 while presenting the general minmax theorems in mixed strategies. Ordinary differential equations appear in Sects. 5.4 (Nash vector fields and dynamics) and 5.5 (Equilibria and evolution). Kolmogorov's extension theorem is applied in Chap. 8, and martingales appear in the last section of this last chapter (Repeated games with incomplete information).

Complementary Reading

We present here several publications that will complement our concise introduction to the mathematical foundations of the theory of strategic games.

An unavoidable encyclopedia in which each chapter covers a specific domain is:

- Aumann R.J. and S. Hart, eds., *Handbook of Game Theory I, II, III*, North Holland, 1992, 1994, 2002.
- Young P. and S. Zamir, eds., *Handbook of Game Theory IV*, North Holland, 2015.

An easy to read classic:

- Owen G., *Game Theory* (3rd Edition), Academic Press, 1995.

A quite general presentation of the field is given in:

- Maschler M., Solan E. and S. Zamir, *Game Theory*, Cambridge University Press, 2013.

Three basic books more related to economics are:

- Myerson R., *Game Theory, Analysis of Conflict*, Harvard University Press, 1991.
- Fudenberg D. and J. Tirole, *Game Theory*, M.I.T. Press, 1992.
- Osborne M. and A. Rubinstein, *A Course in Game Theory*, M.I.T. Press, 1994.

Equilibrium and selection are analyzed in:

- van Damme E., *Stability and Perfection of Nash Equilibria*, Springer, 1991

and more recently in:

- Ritzberger K., *Foundations of Non-Cooperative Game Theory*, Oxford University Press, 2002.

The next two references are more mathematically oriented:

- Barron E.N., *Game Theory, An Introduction*, Wiley, 2008.
- Gonzalez-Diaz J., Garcia-Jurado I. and M. Gloria Fiestras-Janeiro, *An Introductory Course on Mathematical Game Theory*, GSM 115, AMS, 2010.

A reference for evolution games is:

- Hofbauer J. and K. Sigmund, *Evolutionary Games and Population Dynamics*, Cambridge U.P., 1998.

A nice presentation of the links with economics/evolution can be found in:

- Weibull J., *Evolutionary Game Theory*, MIT Press, 1995.

A more recent contribution is:

- Sandholm W., *Population Games and Evolutionary Dynamics*, M.I.T. Press, 2010.

The recent field of algorithmic game theory is presented in:

- Nizan N., Roughgarden T., Tardos E. and V. Vazirani, (eds.) *Algorithmic Game Theory*, Cambridge University Press, 2007.
- Karlin A. and Y. Peres, *Game Theory, Alive*, AMS, 2017.

A short and quick introduction to strategic and repeated games (in French) is:

- Laraki R., Renault J. and T. Tomala, *Théorie des Jeux*, X-UPS 2006, Editions de l'Ecole Polytechnique.

An introduction to zero-sum repeated games is given in:

- Sorin S., *A First Course on Zero-Sum Repeated Games*, Springer, 2002

and a general mathematical analysis of the field is presented in:

- Mertens J.F., Sorin S. and S. Zamir, *Repeated Games*, Cambridge University Press, 2015.

Acknowledgements

The authors thank Miquel Oliu-Barton, Tristan Tomala, Cheng Wan, and also Vianney Perchet, Guillaume Vigeral, and Yannick Viossat for their careful reading and pertinent remarks.

Paris, France/Liverpool, UK   Rida Laraki
                                                           rida.laraki@dauphine.fr

Toulouse, France   Jérôme Renault
                                                     jerome.renault@ut-capitole.fr

Paris, France   Sylvain Sorin
                                                         sylvain.sorin@imj-prg.fr

Contents

Chapter 1
Introduction

1.1 Strategic Interaction

Game theory aims to analyze situations involving strategic interaction where several entities (agents, populations, firms, automata, cells,...) have specific characters (actions, prices, codes, genes...) that impact each other. The mathematical analogy goes beyond optimization since an agent does not control all the variables that affect him. In addition, an agent's choice of his own controlled variable has some consequences on the other agents.

Several levels of modelization have been suggested. Let us review them briefly.

1.1.1 Strategic Games

This framework corresponds to the main topic analyzed in this course. One identifies the autonomous structures that interact, called *players*. By autonomous, we mean that their characters, parameters or choices, called *strategies*, are determined (selected, chosen by the players) independently from each other.

A *profile* of strategies (one for each player) induces an *outcome* and each player uses a real-valued *utility function* defined on this space of outcomes. The main issue is then to define and analyze rational behavior in this framework. (See Sect. 1.3.)

1.1.2 Coalitional Games

In this approach, the initial data is still given by a set I of players but one takes into account all feasible subsets $C \subset I$, called *coalitions*, and an *effectivity function*

© Springer Nature Switzerland AG 2019
R. Laraki et al., *Mathematical Foundations of Game Theory*, Universitext,
https://doi.org/10.1007/978-3-030-26646-2_1

associates to each C the subset of outcomes it can achieve. The issue is now to deduce a specific outcome for the set of all players.

This corresponds to a normative or axiomatic point of view which, starting from requirements on equity, power or efficiency, suggests a solution. Among the several important concepts let us mention: von Neumann–Morgestern solutions, the Shapley value, core, bargaining set, nucleolus, etc. Coalitional and strategic models are linked in both directions:

– From strategic to coalitional games: the analysis at the strategic level of the choice of strategies by a coalition allows one to define a characteristic (or effectivity) function and then apply the program of coalitional games.

– From coalitional to strategic games: also called "Nash's program" [152]. Starting from the effectivity function and a solution, one defines strategies and utility functions such that a rational behavior of the players in the corresponding strategic game induces the chosen solution.

1.1.3 Social Choice and Mechanism Design

This field studies, in the framework of a strategic game played by a set of players, the impact of the rules on the final outcome. The main focus moves from the study of the strategic behavior of the players to the analysis of the consequences of the procedure on the play of the game. Related areas are incitations and contract theories.

1.2 Examples

The following examples illustrate the kinds of questions, tools and applications that occur in game theory.

1.2.1 Stable Matchings

Consider two finite families I and J (men/women, workers/firms,...) with the same cardinality such that each element $i \in I$ (resp. $j \in J$) has a complete strict order on J (resp. I). The issue is the existence and characterization of stable matchings (or marriages), namely bijections π from I to J such that there are no couples $(i, \pi(i) = j), (i', \pi(i') = j')$ with both j' preferred by i to j and i preferred to i' by j'. For example, if the problem is to match men with women, a matching is stable if no unmatched couple prefers to be matched together rather than stay with their actual partners. (See Exercise 1.)

1.2.2 A Bargaining Problem

Represent by [0, 1] a set which is to be shared between two players, each of whom have preferences regarding their share of the set. The game evolves in continuous time between times 0 and 1 and the player who stops first (at time t) wins the share [0, t] of the set, his opponent getting the complement. Assume that $a_1(t)$ (resp. $a_2(t)$), describing the evaluation by player 1, resp. player 2, of the share [0, t], is a continuous function increasing from 0 to 1 ($1 - a_i(t)$ is the evaluation of the complement (t, 1]). Each player i can obtain $1/2$ by aiming to stop at time t_i with $a_i(t_i) = 1/2$ (if the opponent stops before, even better). However, if $t_i < t_j$ and player i knows it, he can anticipate that player j will not stop before t_j and try to wait until some $t_j - \varepsilon$. Questions appear related to information on the characteristics of the opponent, to the anticipation of his behavior (rationality) and to the impact of the procedure on the outcome (if $t_i < t_j$ then player j would like the splitting to move from 1 to 0). (See Exercise 2.)

1.2.3 Transport Equilibrium

Represent by the interval [0, 1] a very large set of small players, each of which is using either a car or the tube. Assume that they all have the same evaluation of the traffic, which is represented by an increasing function v (resp. m) from [0, 1] to itself, $v(t)$ (resp. $m(t)$) being their utility if they use a car (resp. the tube) while a proportion $t \in [0, 1]$ of the population takes the tube. If $v > m$ the only equilibrium is $t = 0$, even if the outcome $v(0)$ can be inferior to another feasible outcome $m(1)$. If the curves m and v intersect, the corresponding points are equilibria, which may or may not be stable. (See Exercise 3.)

1.2.4 Auctions

An indivisible object is proposed for auction and n players have valuations v_i, $i = 1, \ldots, n$, upon it. One can consider decreasing auctions where the proposed price decreases until acceptance or increasing auctions where the players make successive increasing offers. Another model corresponds to the case where the players simultaneously provide sealed bids b_i and the item is given to the one with the highest bid. If the price to pay is this highest bid, the players are interested in knowing the valuation of their opponents. If the price to pay corresponds to the second best bid, the strategy $b_i = v_i$ (bidding his true valuation) is dominant for all players. (See Sect. 1.3.2 and Exercise 4.)

1.2.5 Condorcet's Paradox

Three players a, b, c have strict preferences on three candidates A, B, C. If a ranks the candidates $A > B > C$, b ranks $B > C > A$ and c ranks $C > A > B$, then A is preferred to B by a majority, B is preferred to C by a majority, and also C is preferred to A by a majority. Hence, majority rule is not always transitive, and there may be no candidate who can win against all other candidates by majority rule.

We now provide several models of dynamical interactions.

1.2.6 An Evolution Game

Consider a competition between three types of cells: a produces a virus and an anti-virus, b produces an anti-virus and c produces nothing. Production has a cost, hence b wins against a (in terms of fitness or reproduction rate) and c wins against b but a kills c. One faces a cycle. The corresponding dynamics on the simplex of proportions of the types in the population has an interior rest point (where the three types are simultaneously present) but it may be attractive or repulsive. (See Chap. 5.)

1.2.7 A Stochastic Game

Consider a situation where two fishermen fish a species that can be in large quantity (a), or small quantity (b), or in extinction (c). Players have an intensive (I) or reduced (R) activity and the outcome of their activity, which depends on the state of the species (a, b or c), is an amount of fish and a probability distribution on the state for the next period. This defines a *stochastic game*.

In state a the description of the strategic interaction is given by:

	I	R
I	100, 100	120, 60
R	60, 120	80, 80

and the evolution of the state is defined by:

	I	R
I	$(0.3, 0.5, 0.2)$	$(0.5, 0.4, 0.1)$
R	$(0.5, 0.4, 0.1)$	$(0.6, 0.4, 0)$

For example, if in this state a player 1 (whose strategies are represented by the rows of the matrix) fishes a lot (I) and player 2 (represented by the columns) has a reduced activity (R), the evaluation (payoff, utility) of the outcome for player 1 drop is 120, the evaluation is 60 for player 2, and the state the next day will be a with probability 0.5, resp. (b; 0.4) and (c; 0.1).

In state (b) the data for the utilities is:

	I	R
I	50, 50	60, 30
R	30, 60	40, 40

and for the transitions probabilities:

	I	R
I	(0, 0.5, 0.5)	(0.1, 0.6, 0.3)
R	(0.1, 0.6, 0.3)	(0.8, 0.2, 0)

and in state c the fishing is 0 and the state is absorbing, i.e. stays in state c at all future stages, whatever the players' choices.

When choosing an activity, there is clearly a conflict between the immediate outcome (today's payoff) and the consequence on the future states (and hence future payoffs). Thus the evaluation of some behavior today depends upon the duration of the interaction. (See Sect. 8.5.)

1.2.8 A Repeated Game

Consider a repeated game between two players where the stage interaction is described by the payoff matrix:

	α	β
a	10, 0	1, 1
b	5, 5	0, 0

(Player 1 chooses a row a or b, player 2 a column α or β, and the corresponding entry indicates both players' utilities of the induced outcome.) At each stage, the players make choices (independently) and they are informed of both choices before the next stage. If one does not care about the future, one gets a repetition of the profile (a, β) (since a is always better than b and then β is better than α facing a) but one can introduce threats of the type "play β forever in the future" in case of deviation from b to a, to stabilize the repetition of the outcome (b, α). The use of *plan* and *threat* is fundamental in the study of repeated games. (See Chap. 8.)

1.3 Notations and Basic Concepts

We introduce here the main notions needed to analyze games.

1.3.1 Strategic Games

A *strategic game in normal form G* is defined by:

– a finite set I of *players* (with cardinality $|I| = n \geqslant 1$);
– a non-empty set of *strategies* S^i for each player $i \in I$;
– a mapping $g = (g^1, \ldots, g^n)$ from $S = \Pi_{i=1}^{n} S^i$ to $\mathbb{R}^n$.

This modelizes choices made independently by the players (each $i \in I$ can choose a strategy $s^i \in S^i$) and the global impact on player j when the *profile* $s = (s^1, \ldots, s^n)$ is chosen is measured by $g^j(s) = g^j(s^1, \ldots, s^n)$, called the *payoff* of player j, and g^j is the *payoff function* of j.

We will also use the alternative notation $s = (s^i, s^{-i})$, where s^{-i} stands for the vector of strategies $(s^j)_{j \in I \setminus \{i\}}$ of players other than i, and $S^{-i} = \Pi_{j \neq i} S^j$.

More generally, a *game form* is a mapping F from S to an outcome space R. Each player $i \in I$ has a preference (e.g. a total preorder $\succeq_i$) on R. If this is represented by a *utility function* u^i from R to $\mathbb{R}$, then the composition $u^i \circ F$ gives g^i, the payoff evaluation of the outcome by player i. (This amounts to taking as the space of outcomes the set S of strategy profiles.)

Note that $g^i(s)$ is not a physical amount (of money for example) but the utility for player i of the outcome induced by the profile s. By definition player i *strictly prefers s to t* iff $g^i(s) > g^i(t)$ and in the case of a lottery on profiles, we assume that the evaluation of the lottery is the expectation of g^i. Thus, we assume in this book that the players' preferences over lotteries satisfy the von Neumann–Morgenstern axioms.

1.3.2 Domination

Given a set L and x, y in $\mathbb{R}^L$ we write:

$x \geqslant y$ if $x^l \geqslant y^l$ for each $l \in L$,
$x > y$ if $x^l \geqslant y^l$ for each $l \in L$, and $x \neq y$,
$x \gg y$ if $x^l > y^l$ for each $l \in L$.

In a strategic game, s^i in S^i is a *strictly dominant* (resp. *dominant*) strategy of player i if

$$g^i(s^i, \cdot) \gg g^i(t^i, \cdot) \quad \forall t^i \in S^i,$$

resp.

$$g^i(s^i, \cdot) \geqslant g^i(t^i, \cdot) \quad \forall t^i \in S^i.$$

s^i is *strictly dominated* (resp. *weakly dominated*) if there exists a t^i in S^i with

$$g^i(s^i, \cdot) \ll g^i(t^i, \cdot),$$

resp.

$$g^i(s^i, \cdot) < g^i(t^i, \cdot).$$

1.3.3 Iterated Elimination

A strictly dominated strategy is considered a bad choice for a player. Suppose we start from a strategic game and remove all strictly dominated strategies. This elimination defines new sets of strategies, hence new possible strictly dominated strategies, and we can iterate the process by eliminating the new strictly dominated strategies, etc. The game is *solvable* if the *iterated elimination* of strictly dominated strategies converges to a singleton (in particular, if each player has a strictly dominant strategy, see Chap. 4 and the example in Sect. 1.4.1).

1.3.4 Best Response

The *ε-best response correspondence* (or *ε-best reply correspondence*) ($\varepsilon \geqslant 0$), $\mathrm{BR}^i_\varepsilon$ from S^{-i} to S^i, is defined by:

$$\mathrm{BR}^i_\varepsilon(s^{-i}) = \{s^i \in S^i : g^i(s^i, s^{-i}) \geqslant g^i(t^i, s^{-i}) - \varepsilon, \forall t^i \in S^i\}.$$

A *Nash equilibrium* is a profile s of strategies such that no unilateral deviation is profitable: for each player i an alternative choice t^i would not give him a higher payoff: $g^i(t^i, s^{-i}) \leqslant g^i(s)$. This means: $s^i \in \mathrm{BR}^i_0(s^{-i}), \forall i \in I$. (See Chap. 4.)

1.3.5 Mixed Extensions

In the case of measurable strategy spaces S^i (with σ-algebra $\mathcal{S}^i$), a *mixed extension* of a game G is defined by a space of *mixed strategies* M^i, for each $i \in I$, which is a subset of the set of probabilities on $(S^i, \mathcal{S}^i)$, convex and containing S^i (elements of S^i are identified with the Dirac masses and are called *pure strategies*).

We assume that Fubini's theorem applies to the integral of g on $M = \Pi^n_{i=1} M^i$. This allows us to define the extended payoff as the expectation with respect to the product probability distribution generated by the mixed strategies of the players:

$$g^i(m) = \int_S g^i(s) \; m^1(ds^1) \otimes \ldots \otimes m^n(ds^n), \quad m \in M.$$

Explicitly, in the finite case if $\Delta(S^i)$ is the simplex (convex hull) of S^i, $\Delta = \prod_{i=1}^{N} \Delta(S^i)$ denotes the set of mixed strategy profiles. Given the profile $\sigma = (\sigma^1, \ldots, \sigma^N) \in \Delta$, the payoff for player i in the mixed extension of the game is given by

$$g^i(\sigma) = \sum_{s=(s^1,\ldots,s^n)\in S} \left(\prod_j \sigma^j(s^j) \right) g^i(s)$$

and corresponds to the multilinear extension of g^i.

1.4   Information and Rationality

We present here several examples that underline the difference between one-player games (or optimization problems) and strategic interaction.

1.4.1   *Dominant Strategy and Dominated Outcome*

	α	β
a	10, 0	1, 1
b	5, 5	0, 0

As usual in the matrix representation, the convention is that player 1 chooses the row, player 2 the column, and the first (resp. second) component of the matrix entry is the payoff of player 1 (resp. player 2). Here $I = \{1, 2\}$, $S^1 = \{a, b\}$ and $S^2 = \{\alpha, \beta\}$.

In this game a strictly dominates b and β is a best response to a, hence the outcome is $(1, 1)$ by iterated elimination of strictly dominated strategies. Note that if a is not available to player 1, the outcome is $(5, 5)$, which is better for both players.

1.4.2   *Domination and Pareto Optimality*

Given a set of feasible payoffs $S \subset \mathbb{R}^n$, an element x in S is called a Pareto optimum (or Pareto-efficient) if there exists no other element $y \neq x$ in S such that $y_i \geqslant x_i$ for each $i = 1, \ldots, n$.

Consider the following two-player game with three strategies for each player:

1, 0	0, 1	1, 0
0, 1	1, 0	1, 0
0, 1	0, 1	1, 1

In the set $\{(1, 0), (0, 1), (1, 1)\}$ of feasible payoffs, $(1, 1)$ is the only Pareto optimum and is eliminated via weak domination, leading to:

1, 0	0, 1
0, 1	1, 0

where the only rational outcome is $(1/2, 1/2)$.

1.4.3 Order of Elimination

	α	β
a	2, 2	1, 2
b	2, 0	0, 1

The outcome $(2, 2)$ is eliminated by weak domination if one starts with player 2, but not if one starts with player 1. However it is easy to check that there is no ambiguity in the process of iterated elimination via *strict* domination.

1.4.4 Knowledge Hypotheses

It is crucial when analyzing the deductive process of the players to distinguish between:

- knowledge of the state, or *factual knowledge*, which corresponds to information on the game parameters: strategies, payoffs (in this framework the *autonomous procedures* for a player are those that depend only on his own set of strategies and payoff function); and
- knowledge of the world, or *epistemic knowledge*, which means in addition knowledge of the information and the (amount of) rationality of the opponents.

One then faces a circular paradox: to define the rationality of a player, one has to specify his information, which includes, among other things, his information on the rationality of his opponents (see e.g. Sorin [200]).

1.4.5   Domination and Mixed Extension

The notion of domination depends on the framework:

	α	β
a	3	0
b	1	1
c	0	3

Strategy b is not dominated by the pure strategy a or by c but is strictly dominated by the mixed strategy $\frac{1}{2}a + \frac{1}{2}c$.

1.4.6   Dynamic Games and Anticipation

The main difference in dynamic interactive situations between the modelization in terms of repeated or evolution games is whether each player takes into account the consequences of his own present action on the future behavior of the other participants. The first model analyzes strategic interactions involving rational anticipations. The second describes the consequences at the global level of individual adaptive myopic procedures.

1.5   Exercises

Exercise 1. Stable matchings
Consider a set M of n men (denoted by $a, b, c, \ldots$) and a set W of n women (denoted by $A, B, C, \ldots$). Men have strict preferences over women, and women have strict preferences over men. For instance, with $n = 3$, man b may rank woman C first, then A then B, whereas woman C may rank man a first, then c then b. A *matching* is a subset of $M \times W$ such that every man is paired to exactly one woman, and every woman is paired to exactly one man (it can be viewed as a bijection from M to W).

A matching μ is *stable* if there is no couple (x, Y) not paired by μ and such that both x and Y respectively prefer Y and x to their partner under μ (so that x and Y would prefer to "divorce" from their partner given by μ in order to be paired together).

This model was introduced by Gale and Shapley [74], who showed the existence of a stable matching thanks to the following algorithm:

Women stay home and men visit women (another algorithm is obtained by exchanging the roles between M and W).

Day 1: every man proposes to the woman he prefers; if a woman has several offers, she keeps at home the man she likes best and rejects all other propositions. If every woman got exactly one offer, each offer is accepted and the algorithm stops. Otherwise:

Day 2: every man who has been rejected on day 1 proposes to his next (hence second) preferred woman. Then each woman compares her new offers (if any) to the man she kept on day 1 (if any), and keeps at home the man she likes best. If every woman keeps a man, the algorithm stops. Otherwise, inductively:

Day k: every man who was rejected on day $k-1$ proposes to the next woman on his list. Then each woman compares her new offers (if any) to the man she kept at day $k-1$ (if any), and keeps at home the man she likes best. If every woman keeps a man, the algorithm stops. Otherwise, at least one woman and one man are alone at the end of day k, and the algorithm continues.

(1) Show that the algorithm is well defined and stops in at most n^2 days.
(2) Compute the stable matchings for each of the following two preferences:

	A	B	C
a	(1, 3)	(2, 2)	(3, 1)
b	(3, 1)	(1, 3)	(2, 2)
c	(2, 2)	(3, 1)	(1, 3)

and

	A	B	C	D
a	(1, 3)	(2, 3)	(3, 2)	(4, 3)
b	(1, 4)	(4, 1)	(3, 3)	(2, 2)
c	(2, 2)	(1, 4)	(3, 4)	(4, 1)
d	(4, 1)	(2, 2)	(3, 1)	(1, 4)

For example, the entry $(1, 3)$ in position (a, A) means that man a ranks woman A first, and that woman A ranks man a third.

(3) Show that the final outcome of the algorithm is necessarily a stable matching.
(4) Show that a stable matching may not exist in the following variant, where there is a single community of people: there are $2n$ students and n rooms for two, and each student has a strict preference over his $2n-1$ potential room-mates.
(5) Define stable matchings and study their existence in a society with polygamy: consider n students and m schools, each student has preferences over schools, and each school has a ranking over students and a maximal capacity (*numerus clausus*).

Exercise 2. A sharing procedure
A referee cuts a cake, identified with the segment $[0, 1]$, between two players as follows: he moves his knife continuously from 0 to 1. The first player to stop the knife at some $x \in [0, 1]$ receives the piece of cake to the left of the knife, i.e. $[0, x]$, whereas the other player receives the remaining part to the right of the knife. If the two players stop the knife simultaneously, by convention player 1 receives the left part and player 2 the right part. The entire cake is valued 1 by each player. Player 1 (resp. player 2) values the left part of x as $f(x)$ (resp. $g(x)$), and the right part as $1 - f(x)$ (resp. $1 - g(x)$), where f and g are continuous, onto and strictly increasing from $[0, 1]$ to itself.

(1) Show that each player can guarantee to have at least $1/2$.
(2) What should you do if you are player 1 and you know f and g?
(3) Discuss the case where you are player 1 (resp. player 2) and do not know g (resp. f).
(4) The referee changes the game and now moves his knife continuously from right to left (from $x = 1$ to $x = 0$). Are the players indifferent to this modification?

Exercise 3. Bus or car?

Consider a very large city, with a population modeled as the interval $[0, 1]$. Individuals have the choice between taking the bus or their private car, and all have the same preferences: $u(B, x)$ (resp. $u(V, x)$) is the utility of someone taking the bus (resp. his car) whenever x is the proportion of the population taking the bus. We make the following natural assumptions: $u(V, x)$ strictly increases with x (high x implies less traffic) and $u(B, \cdot)$ and $u(V, \cdot)$ are continuous. An initial proportion $x_0 \in (0, 1)$ is given, and the proportion x_t of bus users at time $t \in [0, +\infty[$ is assumed to follow the replicator dynamics

$$\dot{x}_t = (u(B, x_t) - v(x_t)) x_t,$$

with $v(x_t) = x_t u(B, x_t) + (1 - x_t) u(V, x_t)$.

Let us study the dependence of the stationary points of the dynamics on the utility functions.

(1) What happens if $u(V, x) > u(B, x)$ for all x?
(2) Give an example where the "social welfare" $v(x_t)$ strictly decreases with time.
(3) What if $u(V, x) = 2 + 3x$ and $u(B, x) = 3$ for all x?
(4) Assume that $u(B, 0) > u(V, 0)$ and $u(B, 1) < u(V, 1)$. Show that generically, the number m of stationary points of the dynamics (which are limits of trajectories starting in $(0, 1)$) is odd, and that $(m + 1)/2$ of them are local attractors of the dynamics.
(5) Proceed as in (4) in the other cases (always considering generic cases).

Exercise 4. A Vickrey auction

To sell a painting, the following auction is organized:

(a) each player i submits a bid p_i in a sealed envelope;
(b) the player (say k) submitting the highest bid ($p_k = \max_i p_i$) wins the auction and receives the painting;
(c) if several players submit the same highest bid, the winner is chosen according to a given linear order (this order, or tie-breaking rule, will play no rôle for the result; one can also assume that the winner is selected uniformly among the players with the highest bid);
(d) the winner k pays for the painting the second best price p, defined as $p = \max_{i \neq k} p_i$.

Assume that each player i has a given valuation v_i for the painting (v_i is the maximal amount he is willing to pay to obtain the painting), and that his utility is given by 0 if he is not the winner, and $v_i - p$ if he wins the auction and pays p.

Show that for each player i, submitting the bid $p_i = v_i$ is a dominant strategy.

Chapter 2
Zero-Sum Games: The Finite Case

2.1 Introduction

Zero-sum games are two-person games where the players have opposite evaluations of outcomes, hence the sum of the payoff functions is 0. In this kind of strategic interaction the players are antagonists, hence this induces pure conflict and there is no room for cooperation. It is thus enough to consider the payoff of player 1 (which player 2 wants to minimize).

A finite zero sum game is represented by a real-valued matrix A. The first important result in game theory is called the minmax theorem and was proved by von Neumann [224] in 1928. It states that one can associate a number $v(A)$ to this matrix and a way of playing for each player such that each can guarantee this amount. This corresponds to the notions of "value" and "optimal strategies".

This chapter introduces some general notations and concepts that apply to any zero-sum game, and provides various proofs and extensions of the minmax theorem. Some consequences of this result are then given. Finally, a famous learning procedure (fictitious play) is defined and we show that the empirical average of the stage strategies of each player converges to the set of optimal strategies.

2.2 Value and Optimal Strategies

Definition 2.2.1 A *zero-sum game* G in *strategic form* is defined by a triple (I, J, g), where I (resp. J) is the non-empty set of strategies of player 1 (resp. player 2) and $g : I \times J \longrightarrow \mathbb{R}$ is the payoff function of player 1.

The interpretation is as follows: player 1 chooses i in I and player 2 chooses j in J, in an independent way (for instance simultaneously). The payoff of player 1 is then $g(i, j)$ and that of player 2 is $-g(i, j)$: this means that the evaluations of the outcome induced by the joint choice (i, j) are opposite for the two players. Player 1

© Springer Nature Switzerland AG 2019
R. Laraki et al., *Mathematical Foundations of Game Theory*, Universitext,
https://doi.org/10.1007/978-3-030-26646-2_2

wants to maximize g and is called the maximizing player. Player 2 is the minimizing player. With the notations of Chap. 1, the strategy sets are $S^1 = I$ and $S^2 = J$ and the payoff functions are $g^1 = g = -g^2$, hence the terminology "zero-sum". Each of the two players knows the triple (I, J, g).

$G = (I, J, g)$ is a *finite* zero-sum game when I and J are finite. The game is then represented by an $I \times J$ matrix A, where player 1 chooses the row $i \in I$, player 2 chooses the column $j \in J$ and the entry A_{ij} of the matrix corresponds to the payoff $g(i, j)$. A basic example is the game called "Matching Pennies":

1	−1
−1	1

Conversely, any real matrix can be considered as a finite zero-sum game, also called a *matrix game*.

Consider from now on a given zero-sum game $G = (I, J, g)$. Player 1 aims to maximize the payoff function g, but this one depends upon the two variables i and j while player 1 controls only i and not j. On the other hand player 2 seeks to minimize g and controls j but not i.

Definition 2.2.2 Player 1 *guarantees* $w \in \mathbb{R} \cup \{-\infty\}$ in G if there exists a strategy $i \in I$ that induces a payoff of at least w, that is,

$$\exists i \in I, \ \forall j \in J, \quad g(i, j) \geqslant w.$$

Symmetrically, player 2 *guarantees* $w \in \mathbb{R} \cup \{+\infty\}$ in G if there exists a strategy $j \in J$ that induces a payoff of at most w, that is,

$$\exists j \in J, \ \forall i \in I, \quad g(i, j) \leqslant w.$$

It is clear that for each i in I, player 1 (or strategy i) guarantees $\inf_{j \in J} g(i, j)$ and similarly for any j in J, player 2 guarantees $\sup_{i \in I} g(i, j)$.

Definition 2.2.3 The *maxmin* of G, denoted by $\underline{v}$, is given by

$$\underline{v} = \sup_{i \in I} \inf_{j \in J} g(i, j) \in \mathbb{R} \cup \{-\infty, +\infty\}.$$

Similarly, the *minmax* of G, denoted by $\overline{v}$, is

$$\overline{v} = \inf_{j \in J} \sup_{i \in I} g(i, j) \in \mathbb{R} \cup \{-\infty, +\infty\}.$$

Hence the maxmin is the supremum of the quantities that player 1 guarantees and the minmax is the infimum of the quantities that player 2 guarantees. The maxmin can also be considered as the evaluation of the interaction scheme where player 1 would first choose $i \in I$, then player 2 would choose $j \in J$ knowing i. This corresponds to the worst situation for player 1 and thus leads to a lower bound on his payoff in the play of the game. Similarly the minmax is associated to the situation where player 2 plays first, then player 1 plays knowing the strategy of his opponent. (Note that if player 1 plays first but his choice is not known by player 2, the choices are independent.) The fact that the first situation is less favorable to player 1 translates to the next lemma:

Lemma 2.2.4

$$\underline{v} \leqslant \overline{v}.$$

Proof For all $(i, j) \in I \times J$ we have $g(i, j) \geqslant \inf_{j' \in J} g(i, j')$. Taking the supremum w.r.t. i on each side, we get $\sup_{i \in I} g(i, j) \geqslant \underline{v}$, for all j in J. Taking now the infimum w.r.t. j, we obtain $\overline{v} \geqslant \underline{v}$. $\square$

The jump $\overline{v} - \underline{v}$ is called the *duality gap*.

Definition 2.2.5 The game G has a *value* if $\underline{v} = \overline{v}$. In this case the value v of G is denoted by $\text{val}(G) = v = \underline{v} = \overline{v}$.

In the Matching Pennies example one obtains $\underline{v} = -1 < 1 = \overline{v}$ and the game has no value (we will see later on that the mixed extension of the game has a value).

When the game has a value, it corresponds to the rational issue of the interaction in the sense of a fair evaluation, justified by the two (rational) players of the game. The value can then be interpreted as the worth, or the price, of the game G.

Definition 2.2.6 Given $\varepsilon > 0$, a strategy of player 1 is *maxmin ε-optimal* if it guarantees $\underline{v} - \varepsilon$. If the game has a value, we call it more simply *ε-optimal*. 0-optimal strategies are called *optimal*. Dual definitions hold for player 2.

Example 2.2.7 $G = (\mathbb{N}, \mathbb{N}, g)$, where $g(i, j) = 1/(i + j + 1)$. The game has a value and $v = 0$. All strategies of player 1 are optimal, and player 2 has no optimal strategy.

Lemma 2.2.8 *If there exists a w that both players 1 and 2 can guarantee (using i and j respectively), then w is unique and the game has a value, equal to w. Both i and j are optimal strategies.*

Proof We have $w \leqslant \underline{v} \leqslant \overline{v} \leqslant w$. $\square$

When both spaces I and J have a measurable structure, we can introduce the mixed extensions of G (cf. Sect. 1.3.5). If a strategy $i \in I$ guarantees w in G then it guarantees the same amount in any mixed extension (X, Y, g) of G. In fact, the integral being linear w.r.t. y, we have $g(i, y) = \int_J g(i, j) dy(j) \geqslant w$ for all $y \in Y$, as soon as $g(i, j) \geqslant w$ for all $j \in J$.

Hence we deduce:

Lemma 2.2.9 *The duality gap of a mixed extension of G is smaller than the initial duality gap of G.*

In particular, if a zero-sum game has a value, any mixed extension of the game has the same value. In the reminder of this chapter we will mainly consider the case of a finite zero-sum game.

2.3 The Minmax Theorem

In a game, it is natural to let the players choose their strategy in a random way. For example, if we are playing Matching Pennies, or describing an algorithm that will play it "online", it is clearly interesting to select each strategy with probability 1/2, in order to hide our choice from the opponent. An alternative interpretation of "mixed strategies" is to consider the probability assigned to a strategy as his opponent's degree of belief that this strategy will be adopted ([92]: see Chap. 7).

Mathematically, considering mixed strategies allows to deal with convex sets. If S is a finite set of cardinality n, we denote by $\Delta(S)$ the set of probabilities on S or equivalently the simplex on S:

$$\Delta(S) = \{x \in \mathbb{R}^n : x^s \geq 0, s \in S; \textstyle\sum_{s \in S} x^s = 1\}.$$

The mixed extension of a finite game $G = (I, J, g)$ is thus the game $\Gamma = (\Delta(I), \Delta(J), g)$, where the payoff function g is extended in a multilinear way:

$$g(x, y) = \mathbb{E}_{x \otimes y}\, g = \sum_{i,j} x^i y^j g(i, j).$$

An element x of $\Delta(I)$, resp. y of $\Delta(J)$, is a *mixed strategy* of player 1, resp. player 2, in the game Γ. An element $i \in I$ is also considered as a Dirac measure or a unit vector e^i in $\Delta(I)$ and is a *pure strategy* of player 1 in Γ, and similarly for player 2.

The *support* of a mixed strategy x of player 1, denoted supp (x), is the subset of actions $i \in I$ such that $x^i > 0$.

When the game G is represented by a matrix A, an element $x \in \Delta(I)$ will correspond to a row matrix and an element $y \in \Delta(J)$ to a column matrix, so that the payoff can be written as the bilinear form $g(x, y) = xAy$.

Theorem 2.3.1 (Minmax [224]) *Let A be an $I \times J$ real matrix. Then there exist (x^*, y^*, v) in $\Delta(I) \times \Delta(J) \times \mathbb{R}$ such that*

$$x^* Ay \geq v, \quad \forall y \in \Delta(J) \quad \text{and} \quad xAy^* \leq v, \quad \forall x \in \Delta(I). \tag{2.1}$$

In other words, the mixed extension of a matrix game has a value (one also says that any finite zero-sum game has a value in mixed strategies) and both players have optimal strategies.

The real number v in the theorem is uniquely determined and corresponds to the value of the matrix A:

$$v = \max_{x \in \Delta(I)} \min_{y \in \Delta(J)} x A y = \min_{y \in \Delta(J)} \max_{x \in \Delta(I)} x A y,$$

also denoted $\mathtt{val}(A)$.

As a mapping defined on matrices, from $\mathbb{R}^{I \times J}$ to $\mathbb{R}$, the operator $\mathtt{val}$ is continuous, monotonic (increasing) and non-expensive: $|\mathtt{val}(A) - \mathtt{val}(B)| \leqslant \|A - B\|_\infty$ (see Sect. 3.4).

Several proofs of the minmax theorem are available. A classical proof relies on the duality theorem in linear programming in finite dimensions (maximization of a linear form under linear constraints).

Theorem 2.3.2 *Let A be an $m \times n$ matrix, b an $m \times 1$ vector and c a $1 \times n$ vector with real coefficients. The two dual linear programs*

$$\begin{array}{cc} \min\langle c, x \rangle & \max\langle y, b \rangle \\ (\mathcal{P}_1) \quad Ax \geqslant b & (\mathcal{P}_2) \quad yA \leqslant c \\ x \geqslant 0 & y \geqslant 0 \end{array}$$

have the same value as soon as they are feasible, i.e. when the sets $\{Ax \geqslant b; x \geqslant 0\}$ and $\{yA \leqslant c; y \geqslant 0\}$ are non-empty.

Proof This result itself is a consequence of the "alternative theorem" for linear systems, see Exercises 7–9. □

Proof (of the minmax theorem) By considering $A + tE$ with $t \geqslant 0$ and E being the matrix with $E_{ij} = 1, \forall (i, j) \in I \times J$, one can assume $A \gg 0$, of dimension $m \times n$.

Let us consider the dual programs

$$\begin{array}{cc} \min\langle X, c \rangle & \max\langle b, Y \rangle \\ (\mathcal{P}_1) \quad XA \geqslant b & (\mathcal{P}_2) \quad AY \leqslant c \\ X \geqslant 0 & Y \geqslant 0 \end{array}$$

where the variables satisfy $X \in \mathbb{R}^m, Y \in \mathbb{R}^n$ and the parameters are given by $c \in \mathbb{R}^m, c_i = 1, \forall i$ and $b \in \mathbb{R}^n, b_j = 1, \forall j$.

$(\mathcal{P}_2)$ is feasible with $Y = 0$, as is $(\mathcal{P}_1)$ by taking X large enough, using the hypothesis on A. Thus by the duality theorem there exists a triple (X^*, Y^*, w) with:

$$X^* \geqslant 0, \ Y^* \geqslant 0, \ X^* A \geqslant b, \ AY^* \leqslant c, \quad \sum_i X_i^* = \sum_j Y_j^* = w.$$

$X^* \neq 0$ implies $w > 0$, hence dividing X^* and Y^* by w, one obtains the existence of $(x^*, y^*) \in \Delta(I) \times \Delta(J)$ with

$$x^* A e^j \geqslant 1/w, \forall j, \quad e^i A y^* \leqslant 1/w, \forall i.$$

Hence there is a value, namely $1/w$, and x^* and y^* are optimal strategies. □

A more constructive proof of von Neumann's minmax theorem can be obtained through an approachability algorithm (see Exercise 4). Alternatively, one can use Loomis' theorem (Theorem 2.5.1), the proof of which uses a recursive argument on the dimension of the strategy sets.

Let us mention that von Neumann's minmax theorem can be generalized to the case where the payoffs are not real numbers but belong to an ordered field (and then the value belongs to the field, [228]): it is enough to check that the set defined through Eq. (2.1) is specified by a finite number of weak linear inequalities, and this is the advantage of the previous "elementary proof" based on Fourier elimination, see Exercise 7.

2.4 Properties of the Set of Optimal Strategies

Consider a matrix game defined by $A \in \mathbb{R}^{I \times J}$. Denote by $X(A)$ (resp. $Y(A)$) the subset of $\Delta(I)$ (resp. $\Delta(J)$) composed of optimal strategies of player 1 (resp. 2). Recall that a *polytope* is the convex hull of finitely many points (which is equivalent, in finite dimensions, to a bounded set formed by the intersection of a finite number of closed half spaces).

Proposition 2.4.1 (a) $X(A)$ and $Y(A)$ are non-empty polytopes.
(b) If $x \in X(A), y \in Y(A), i \in \text{supp}(x)$ and $j \in \text{supp}(y)$, then $iAy = v$ and $xAj = v$ (complementarity).
(c) There exists a pair of optimal strategies (x^*, y^*) in $X(A) \times Y(A)$ satisfying the property of strong complementarity:

$$\forall i \in I, \ \left(x^{*i} > 0 \iff e^i A y^* = v \right) \ \text{and}$$

$$\forall j \in J, \ \left(y^{*j} > 0 \iff x^* A e^j = v \right).$$

(d) $X(A) \times Y(A)$ is the set of saddle points of A, namely elements $(x^*, y^*) \in \Delta(I) \times \Delta(J)$ such that

$$x A y^* \leqslant x^* A y^* \leqslant x^* A y \quad \forall (x, y) \in \Delta(I) \times \Delta(J).$$

Proof The proofs of (a), (b), and (d) are elementary consequences of the definitions and of the minmax theorem.

In fact, property (d) holds for any zero-sum game: it expresses the identity between optimal strategies and *Nash equilibria* (see Chap. 4) for a zero-sum game.

Assertion (c) corresponds to strong complementarity in linear programming (see Exercise 11). A pure strategy is in the support of an optimal mixed strategy of a player iff it is a best response to any optimal strategy of the opponent.   □

2.5 Loomis and Ville's Theorems

The next extension of von Neumann's minmax theorem is due to Loomis.

Theorem 2.5.1 ([122]) *Let A and B be two I × J real matrices, with B ≫ 0. Then there exist (x, y, v) in $\Delta(I) \times \Delta(J) \times \mathbb{R}$ such that*

$$xA \geqslant v \, xB \quad \text{and} \quad Ay \leqslant v \, By.$$

With $B_{ij} = 1$ for all $(i, j) \in I \times J$, one recovers von Neumann's theorem.

Conversely, one can give an elementary proof of Loomis' theorem assuming von Neumann's theorem: the mapping $t \in \mathbb{R} \mapsto \text{val}(A - tB)$ is continuous, strictly decreasing and has limit $+\infty$ at $-\infty$, and $-\infty$ at $+\infty$. Thus there exists a real v such that $\text{val}(A - vB) = 0$, which gives the result.

See Exercise 1 for a direct proof of Loomis' theorem.

An example of use of von Neumann's theorem is the following:

Corollary 2.5.2 *Any stochastic matrix has an invariant distribution.*

Proof Let A be a *stochastic matrix* in $\mathbb{R}^{I \times J}$, that is, satisfying: $I = J$, $A_{ij} \geqslant 0$ and $\sum_{j \in I} A_{ij} = 1, \forall i \in I$. Let $B = A - \text{Id}$, where Id is the identity matrix, and consider the game defined by the matrix B.

The uniform strategy y^* of player 2 guarantees 0, thus the value of B is non-positive. Moreover, against any mixed strategy y of player 2, by choosing a row i with $y^i = \min_{j \in J} y^j$ player 1 gets a non-negative payoff $\sum_j A_{ij} y^j - y^i$, hence $\overline{v}(B) \geqslant 0$. As a consequence, the value of B is zero.

An optimal strategy x^* of player 1 in B satisfies $x^*A - x^* \geqslant 0$, hence, considering the (right) product by y^*, one obtains equality on each component so that $x^*A = x^*$ (or use complementarity). □

Von Neumann's theorem allows us to prove the existence of a value in the following case of a game with a continuum of strategies and continuous payoff, where $\Delta([0, 1])$ is the set of Borel probabilities on $[0, 1]$.

Theorem 2.5.3 ([222]) *Let $I = J = [0, 1]$ and f be a real-valued continuous function on $I \times J$. The mixed extension $(\Delta(I), \Delta(J), f)$ has a value and each player has an optimal strategy. Moreover, for every $\varepsilon > 0$, each player has an ε-optimal strategy with finite support.*

(In particular any mixed extension of the game (I, J, f) has the same value.)

See Exercise 3 for the proof: one considers finer and finer discretizations of the square $[0, 1] \times [0, 1]$ inducing a sequence of finite games, and one extracts a weakly convergent subsequence of optimal strategies. Exercise 7 of Chap. 3 shows that a continuity hypothesis on f is needed in Ville's theorem. However, although joint continuity is not required (see the next Chap. 3), it allows for strong approximation.

2.6 Examples

Example 2.6.1

1	−2
−1	3

Here $v = 1/7$. Player 1 has a unique optimal strategy: play Top with probability 4/7 and Bottom with probability 3/7. Similarly player 2 has a unique optimal strategy: (5/7, 2/7) on (Left, Right).

Example 2.6.2

1	2
0	t

For all $t \in \mathbb{R}$, the game has value $v = 1$, and each player has a unique optimal strategy, which is pure: Top for player 1, Left for player 2.

Example 2.6.3

a	b
c	d

In the case where each player has two actions, either there exists a pair of pure optimal strategies (and then the value is one of the numbers a, b, c, d) or the optimal strategies are completely mixed and the value is given by

$$v = \frac{ad - bc}{a + d - b - c}.$$

Example 2.6.4

1	0
0	t

For $t \in \mathbb{R}^-$, the game has the value $v = 0$.

For $t \in \mathbb{R}^+$, $v = \frac{t}{1+t}$ and the only optimal strategy of player 1 puts a (decreasing) weight $\frac{1}{1+t}$ on the Bottom line: this plays the rôle of a threat, forcing player 2 to play Left with a high probability.

2.7 Fictitious Play

Let A be an $I \times J$ real matrix. The following process, called *fictitious play*, was introduced by Brown [33]. Consider two players repeatedly playing the matrix game A. At each stage n, $n = 1, 2, \ldots$, each player is aware of the previous strategy of his opponent and computes the empirical distribution of the strategies used in the past. He then plays a strategy which is a best response to this average.

Explicitly, starting with any (i_1, j_1) in $I \times J$, consider at each stage n, $x_n = \frac{1}{n} \sum_{t=1}^{n} i_t$, viewed as an element of $\Delta(I)$, and similarly $y_n = \frac{1}{n} \sum_{t=1}^{n} j_t$ in $\Delta(J)$.

Definition 2.7.1 A sequence $(i_n, j_n)_{n \geqslant 1}$ with values in $I \times J$ is the *realization of a fictitious play process for the matrix A* if, for each $n \geqslant 1$, i_{n+1} is a best response of player 1 to y_n for A:

$$i_{n+1} \in \mathrm{BR}^1(y_n) = \{i \in I : e^i A y_n \geqslant e^k A y_n, \forall k \in I\}$$

and j_{n+1} is a best response of player 2 to x_n for A ($j_{n+1} \in \mathrm{BR}^2(x_n)$, defined in a dual way).

The main properties of this procedure are given in the next result.

Theorem 2.7.2 ([177]) *Let $(i_n, j_n)_{n \geqslant 1}$ be the realization of a fictitious play process for the matrix A. Then:*

(1) The distance from (x_n, y_n) to the set of optimal strategies $X(A) \times Y(A)$ goes to 0 as $n \to \infty$. Explicitly: $\forall \varepsilon > 0, \exists N, \forall n \geqslant N, \forall x \in \Delta(I), \forall y \in \Delta(J)$

$$x_n A y \geqslant \mathrm{val}(A) - \varepsilon \quad \text{and} \quad x A y_n \leqslant \mathrm{val}(A) + \varepsilon.$$

(2) The average payoff on the trajectory, namely $\frac{1}{n} \sum_{t=1}^{n} A_{i_t, j_t}$, converges to $\mathrm{val}(A)$.

We will not prove the above theorem here (see Exercise 12) but instead provide an illustration in the continuous time setting. Take as variables the empirical frequencies x_n and y_n, so that the discrete dynamics for player 1 reads as

$$x_{n+1} = \frac{1}{n+1}[i_{n+1} + n x_n] \quad \text{with} \quad i_{n+1} \in \mathrm{BR}^1(y_n)$$

and hence satisfies

$$x_{n+1} - x_n \in \frac{1}{n+1}[\mathrm{BR}^1(y_n) - x_n].$$

The corresponding system in continuous time is now

$$\dot{x}(t) \in \frac{1}{t}\left[\mathrm{BR}^1(y(t)) - x(t)\right].$$

This is a differential inclusion which defines, with a similar condition for player 2, the process CFP: *continuous fictitious play*. We assume the existence of such a CFP process.

Proposition 2.7.3 ([90, 101]) *For the CFP process, the duality gap converges to 0 at a speed $O(1/t)$.*

Proof Make the time change $z(t) = x(\exp(t))$, which leads to the autonomous differential inclusion

$$\dot{x}(t) \in \left[\mathrm{BR}^1(y(t)) - x(t)\right], \quad \dot{y}(t) \in \left[\mathrm{BR}^2(x(t)) - y(t)\right]$$

known as the *best response dynamics* [78].

Write the payoff as $g(x, y) = x A y$ and for $(x, y) \in \Delta(I) \times \Delta(J)$, let

$$L(y) = \max_{x' \in \Delta(I)} g(x', y) \quad M(x) = \min_{y' \in \Delta(J)} g(x, y').$$

Thus the duality gap at (x, y) is defined as $W(x, y) = L(y) - M(x) \geqslant 0$ and the pair (x, y) defines optimal strategies in A if and only if $W(x, y) = 0$.

Let now $(x(t), y(t))_{t \geqslant 0}$ be a solution of CFP. Denote by

$$w(t) = W(x(t), y(t))$$

the evaluation of the duality gap on the trajectory, and write

$$\alpha(t) = x(t) + \dot{x}(t) \in \mathrm{BR}^1(y(t)) \quad \text{and} \quad \beta(t) = y(t) + \dot{y}(t) \in \mathrm{BR}^2(x(t)).$$

We have $L(y(t)) = g(\alpha(t), y(t))$, thus

$$\frac{d}{dt} L(y(t)) = \dot{\alpha}(t) D_1 g(\alpha(t), y(t)) + \dot{y}(t) D_2 g(\alpha(y), y(t)).$$

The envelope theorem (see e.g., Mas-Colell, Whinston and Green [129, p. 964]) shows that the first term collapses and the second term is $g(\alpha(t), \dot{y}(t))$ (since g is linear w.r.t. the second variable). Then we obtain

$$\begin{aligned}
\dot{w}(t) &= \frac{d}{dt} L(y(t)) - \frac{d}{dt} M(x(t)) \\
&= g(\alpha(t), \dot{y}(t)) - g(\dot{x}(t), \beta(t)) \\
&= g(x(t), \dot{y}(t)) - g(\dot{x}(t), y(t)) \\
&= g(x(t), \beta(t)) - g(\alpha(t), y(t)) \\
&= M(x(t)) - L(y(t)) \\
&= -w(t).
\end{aligned}$$

Thus $w(t) = w(0) e^{-t}$. There is convergence of $w(t)$ to 0 at exponential speed, hence convergence to 0 at a speed $O(1/t)$ in the original problem before the time change. The convergence to 0 of the duality gap implies by uniform continuity the convergence of $(x(t), y(t))$ to the set of optimal strategies $X(A) \times Y(A)$. □

Let us remark that by compactness of the sets of mixed strategies, we obtain the existence of optimal strategies in the matrix game (accumulation points of the trajectories). This provides an alternative proof of the minmax theorem, starting from the existence of a solution to CFP.

The result is actually stronger: the set $X(A) \times Y(A)$ is a global attractor for the best response dynamics, which implies the convergence of the discrete time version, hence of the fictitious play process ([101]), i.e. part (1) of Theorem 2.7.2.

We finally prove part (2) of Theorem 2.7.2.

Proof Let us consider the sum of the realized payoffs: $R_n = \sum_{p=1}^{n} g(i_p, j_p)$.
Writing $U_m^i = \sum_{k=1}^{m} g(i, j_k)$, we obtain

$$R_n = \sum_{p=1}^{n} (U_p^{i_p} - U_{p-1}^{i_p})$$

$$= \sum_{p=1}^{n} U_p^{i_p} - \sum_{p=1}^{n-1} U_p^{i_{p+1}}$$

$$= U_n^{i_n} + \sum_{p=1}^{n-1} (U_p^{i_p} - U_p^{i_{p+1}}),$$

but the fictitious play property implies that

$$U_p^{i_p} - U_p^{i_{p+1}} \leqslant 0.$$

Hence $\limsup_{n \to \infty} \frac{R_n}{n} \leqslant \limsup_{n \to \infty} \max_i \frac{U_n^i}{n} \leqslant \mathrm{val}(A)$,   since $\frac{U_n^i}{n} = g(i, y_n)$
$\leqslant \mathrm{val}(A) + \varepsilon$ for n large enough by part 1) of Theorem 2.7.2. The dual inequality
thus implies the result. $\qquad\square$

Note that part (1) and part (2) of Theorem 2.7.2 are independent. In general, convergence of the average marginal trajectories on moves does not imply the property
of average payoff on the trajectory (for example, in Matching Pennies convergence
of the average strategies to $(1/2, 1/2)$ is compatible with a sequence of payoffs 1 or
-1).

2.8 Exercises

Exercise 1. Loomis' theorem [122]
Let A and B be two $I \times J$ matrices, with $B \gg 0$. We will prove) the existence of a
unique $v \in \mathbb{R}$, and elements $s \in \Delta(I), t \in \Delta(J)$, such that

$$sA \geqslant vsB, \qquad At \leqslant vBt.$$

The proof does not use von Neumann's theorem and is obtained by induction on the
dimension $|I| + |J| = m + n$. The result is clear for $m = n = 1$.

(1) Assume the result is true for $m + n - 1$. Define $\lambda_0 = \sup\{\lambda \in \mathbb{R}, \exists s \in \Delta(I),$
$sA \geqslant \lambda sB\}$ and $\mu_0 = \inf\{\mu \in \mathbb{R}, \exists t \in \Delta(J), At \leqslant \mu Bt\}$.

(a) Show that both sup and inf are reached and that $\lambda_0 \leqslant \mu_0$.

If $\lambda_0 = \mu_0$, the result is achieved, hence we assume from now on that $\lambda_0 < \mu_0$.
Let s_0 and t_0 be such that $s_0 A \geqslant \lambda_0 s_0 B$ and $At_0 \leqslant \mu_0 Bt_0$.

(b) Show that $s_0 A - \lambda_0 s_0 B = 0$ and $A t_0 - \mu_0 B t_0 = 0$ cannot both hold.

(c) Assume then that $j \in J$ is such that $s_0 A_j > \lambda_0 s_0 B_j$ (A_j stands for column j of A, and likewise for B_j) and let $J' = J \setminus \{j\}$.

Using the induction hypothesis, introduce $v' \in \mathbb{R}$ and $s' \in \Delta(I)$ associated to the $I \times J'$ submatrices A' of A and B' of B, with $s' A' \geq v' s' B'$. Show that $v' \geq \mu_0 > \lambda_0$, then obtain a contradiction in the definition of λ_0 by using s_0 and s'.

(2) Application: Let B be a square matrix with positive entries. Show that there exists an eigenvector associated to a positive eigenvalue with positive components (Perron–Frobenius).

Exercise 2. minmax with three players

Consider the three-player game where player 1 chooses a probability $x \in \Delta(\{T, B\})$, player 2 chooses $y \in \Delta(\{L, R\})$ and player 3 chooses $z \in \Delta(\{W, E\})$. The probabilities are independent and the payoff is the expectation of the following function g:

	L	R
T	1	0
B	0	0

W

	L	R
T	0	0
B	0	1

E

Compare

$$\max_{(x,y)} \min_z g(x, y, z) \quad \text{and} \quad \min_z \max_{(x,y)} g(x, y, z).$$

Exercise 3. Ville's theorem [222]

Let $X = Y = [0, 1]$ endowed with its Borel σ-field $\mathcal{B}$ and f be a continuous function from $X \times Y$ to $\mathbb{R}$. Consider the zero-sum game G where player 1 chooses σ in $\Delta(X)$ (probability on $(X, \mathcal{B})$), player 2 chooses τ in $\Delta(Y)$, and the payoff is given by

$$f(\sigma, \tau) = \int_{X \times Y} f(x, y) \, d\sigma(x) \, d\tau(y).$$

For each $n \geq 1$, define the matrix game G_n where player 1 chooses an action i in $X_n = \{0, \ldots, 2^n\}$, player 2 chooses an action j in $Y_n = X_n$ and the payoff is given by $G_n(i, j) = f(\frac{i}{2^n}, \frac{j}{2^n})$. Denote by v_n the value (in mixed strategies) of G_n.

(1) Show that in G player 1 can guarantee the amount $\limsup_n v_n$ (up to ε, for any $\varepsilon > 0$). Deduce that G has a value.

(2) Show that each player has an optimal strategy in G. (Represent optimal strategies of $\{G_n\}$ as elements in $\Delta([0, 1])$ and extract a weakly convergent subsequence.)

Exercise 4. Deterministic approachability and minmax

Let C be a non-empty closed convex subset of $\mathbb{R}^k$ (endowed with the Euclidean norm) and $\{x_n\}$ a bounded sequence in $\mathbb{R}^k$.

For $x \in \mathbb{R}^k$, $\Pi_C(x)$ stands for the closest point to x in C (which is also the projection of x on C) and $\bar{x}_n$ is the Cesàro mean up to stage n of the sequence $\{x_i\}$:

$$\bar{x}_n = \frac{1}{n} \sum_{i=1}^{n} x_i.$$

(1) Deterministic approachability (Blackwell [27]). $\{x_n\}$ is a Blackwell C-sequence if it satisfies

$$\langle x_{n+1} - \Pi_C(\bar{x}_n), \bar{x}_n - \Pi_C(\bar{x}_n) \rangle \leqslant 0, \quad \forall n.$$

Show that it implies that $d_n = d(\bar{x}_n, C)$ converges to 0.

(2) Consequence: the minmax theorem. Let A be an $I \times J$ matrix and assume that the minmax is 0:

$$\bar{v} = \min_{t \in \Delta(I)} \max_{i \in I} e^i At = 0.$$

We will show that player 1 can guarantee 0, i.e. $\underline{v} \geqslant 0$. Let us define by induction a sequence $x_n \in \mathbb{R}^k$, where $k = |J|$. The first term x_1 is any row of the matrix A. Given $x_1, x_2, \ldots, x_n$, define x_{n+1} as follows:
Let $\bar{x}_n^+$ be the vector with j^{th} coordinate equal to $\max(\bar{x}_n^j, 0)$. If $\bar{x}_n = \bar{x}_n^+$ then choose x_{n+1} as any row of A. Otherwise let $a > 0$ such that

$$t_{n+1} = \frac{\bar{x}_n^+ - \bar{x}_n}{a} \in \Delta(J).$$

Since $\bar{v} = 0$, there exists an $i_{n+1} \in I$ such that $e^{i_{n+1}} At_{n+1} \geqslant 0$. x_{n+1} is then the i_{n+1}th row of the matrix A.

(a) Show that $\{x_n\}$ is a Blackwell C-sequence with $C = \{x \in \mathbb{R}^k; \ x \geqslant 0\}$.
(b) Conclude the existence of $s \in \Delta(I)$ with $s At \geqslant 0$, for all $t \in \Delta(J)$.

Exercise 5. Computation of values
Compute the value and the optimal strategies in the following game:

	b_1	b_2
a_1	3	−1
a_2	0	0
a_3	−2	1

Exercise 6. A diagonal game
Compute the value and optimal strategies for the diagonal $n \times n$ game with $a_i > 0$ for all $i = 1, \ldots, n$, given by

$$\begin{pmatrix} a_1 & 0 & \ldots & 0 \\ 0 & a_2 & \ldots & 0 \\ \ldots & \ldots & \ldots & \ldots \\ 0 & 0 & \ldots & a_n \end{pmatrix}$$

Exercise 7. The theorem of the alternative

*Let A be an $m \times n$ matrix and b an $m \times 1$ vector with real components.
Define*

$$S = \{x \in \mathbb{R}^n; \, Ax \geqslant b\}, \qquad T = \{u \in \mathbb{R}^m; \, u \geqslant 0, \, uA = 0, \, \langle u, b \rangle > 0\}.$$

Then one and only one of the two sets S and T is non-empty.

(1) Prove that "S and T non-empty" is impossible.

The proof that $\{S = \varnothing \Rightarrow T \neq \varnothing\}$ is by induction on the number of effective variables (i.e. with a non-zero coefficient) in S, say n.

(2) Prove the assertion for $n = 0$.

(3) Define $I = L \cup K \cup I_0$ with $L \cup K \neq \varnothing$, where $A_{\ell,n+1} > 0, \, \forall \ell \in L, \, A_{k,n+1} < 0, \, \forall k \in K$ and $A_{i,n+1} = 0, \, \forall i \in I_0$.

(a) If $K = \varnothing$, S can be written as

$$\sum_{j=1}^{n} A_{i,j} x_j \geqslant b_i, \qquad i \in I_0,$$

$$x_{n+1} \geqslant \frac{b_\ell}{A_{\ell,n+1}} - \sum_{j=1}^{n} \frac{A_{\ell,j}}{A_{\ell,n+1}} x_j, \qquad \ell \in L.$$

Show that $S = \varnothing$ implies that S_0 (defined by the first family with $i \in I_0$ above) is also empty. Deduce that

$$T_0 = \{\overline{u} \in \mathbb{R}^r; \, \overline{u}\overline{A} = 0, \, \langle \overline{u}, \overline{b} \rangle > 0\}$$

(where $\overline{A}$ is the restriction of A to I_0 and similarly for $\overline{b}$) is non-empty, and finally $T \neq \varnothing$.

(b) If both L and K are non-empty, S reads as

$$\sum_{j=1}^{n} A_{i,j} x_j \geqslant b_i, \qquad i \in I_0,$$

$$x_{n+1} \geqslant \frac{b_\ell}{A_{\ell,n+1}} - \sum_{j=1}^{n} \frac{A_{\ell,j}}{A_{\ell,n+1}} x_j, \qquad \ell \in L,$$

$$x_{n+1} \leqslant \frac{b_k}{A_{k,n+1}} - \sum_{j=1}^{n} \frac{A_{k,j}}{A_{k,n+1}} x_j, \qquad k \in K.$$

Show that $S = \varnothing$ implies that the following set, involving n variables, is empty:

$$\sum_{j=1}^{n} A_{i,j} x_j \geqslant b_i, \qquad i \in I_0,$$

$$\sum_{j=1}^{n} \left(\frac{A_{\ell,j}}{A_{\ell,n+1}} - \frac{A_{k,j}}{A_{k,n+1}} \right) x_j \geqslant \frac{b_\ell}{A_{\ell,n+1}} - \frac{b_k}{A_{k,n+1}}, \qquad (\ell, k) \in L \times K.$$

Deduce, using the induction hypothesis, that the following set is non-empty:

$$v_i \geqslant 0, \qquad i \in I_0; \qquad v_{\ell,k} \geqslant 0, \qquad (\ell, k) \in L \times K,$$

$$\sum_{i \in I_0} A_{ij} v_i + \sum_{(\ell,k) \in L \times K} \left(\frac{A_{\ell,j}}{A_{\ell,n+1}} - \frac{A_{k,j}}{A_{k,n+1}} \right) v_{\ell,k} = 0,$$

$$j = 1, \dots, n$$

$$\sum_{i \in I_0} b_i v_i + \sum_{(\ell,k) \in L \times K} \left(\frac{b_\ell}{A_{\ell,n+1}} - \frac{b_k}{A_{k,n+1}} \right) v_{\ell,k} > 0.$$

Then define u by

$$u_i = v_i, \quad i \in I_0, \quad u_\ell = \sum_{k} \frac{v_{\ell,k}}{A_{\ell,n+1}}, \quad \ell \in L \quad \text{and}$$

$$u_k = \sum_{\ell} -\frac{v_{\ell,k}}{A_{k,n+1}}, \quad k \in K,$$

and prove that $u \in T$.

Comments The proof (*Fourier's elimination method*) provides an explicit algorithm to compute a point in S if this set is non-empty, and otherwise a point in T. It shows explicitly that if the coefficients are in an ordered field (for example $\mathbb{Q}$), then there exists a solution with coefficients in the field.

Exercise 8. Farkas' lemma
Keep the notations of the previous exercise. In addition: Let c be a $1 \times n$ vector, $d \in \mathbb{R}$ and assume $S = \{x \in \mathbb{R}^n, Ax \geqslant b\}$ is non-empty. Then

(i) $\forall x \in \mathbb{R}^n, (Ax \geqslant b \Rightarrow \langle c, x \rangle \geqslant d)$

is equivalent to:

(ii) $\quad \exists u \in \mathbb{R}^m, u \geqslant 0, uA = c$ and $\langle u, b \rangle \geqslant d$.

(1) (ii) implies (i) is clear.
(2) Assume $U = \{u \in \mathbb{R}^m, u \geqslant 0, uA = c$ and $\langle u, b \rangle \geqslant d\}$ is empty. Write it as

$$u[I, A, -A, b] \geqslant [0, c, -c, d],$$

and use the alternative theorem (Exercise 7) to obtain the existence of (p, q, r, t) $\in \mathbb{R}^m_+ \times \mathbb{R}^n_+ \times \mathbb{R}^n_+ \times \mathbb{R}_+$ with

$$p + A(q - r) + tb = 0, \qquad \langle c, q - r \rangle + td > 0.$$

Finally, construct x contradicting $i)$ (if $t = 0$ use $S \neq \varnothing$).

Exercise 9. Duality in linear programming

A *linear program* $\mathcal{P}$ is given by:

$$\begin{aligned} &\min \quad \langle c, x \rangle \\ (\mathcal{P}) \ &Ax \geqslant b \\ &x \geqslant 0 \end{aligned}$$

and $v(\mathcal{P})$ is its value.

$\mathcal{P}$ is *feasible* if $R = \{Ax \geqslant b, x \geqslant 0\} \neq \varnothing$ and *bounded* if $\exists k \in \mathbb{R}, \langle c, x \rangle \geqslant k, \forall x \in R$.

x^* is an *optimal solution* if $x^* \in R$ and $x \in R$ implies $\langle c, x \rangle \geqslant \langle c, x^* \rangle$.

(1) *$\mathcal{P}$ is feasible and bounded if and only if it has an optimal solution.*

The "if" part is clear. For the reverse implication, let $k = \inf_{x \in R} \langle c, x \rangle$ and assume $W = \{x \in \mathbb{R}^n; Ax \geqslant b, x \geqslant 0, -\langle c, x \rangle \geqslant -k\}$ is empty. Using the alternative theorem (Exercise 7) some associated set Z is non-empty. Show that for $\varepsilon > 0$ small enough, Z_ε corresponding to $k + \varepsilon$ is still non-empty, hence (again by the alternative theorem) W_ε is empty, contradicting the definition of k.

The *dual* $\mathcal{P}^*$ of $\mathcal{P}$ is defined by:

$$\begin{aligned} &\max \quad \langle u, b \rangle \\ (\mathcal{P}^*) \ &uA \leqslant c \\ &u \geqslant 0 \end{aligned}$$

(Note that the dual of $\mathcal{P}^*$ is $\mathcal{P}$.)

(2) *$\mathcal{P}$ has an optimal solution if and only if $\mathcal{P}^*$ has one. Moreover, in this case both programs have the same value.*

Let $\bar{x}$ be an optimal solution so that

$$Ax \geqslant b, \quad x \geqslant 0 \Longrightarrow \langle c, x \rangle \geqslant \langle c, \bar{x} \rangle.$$

Use Farkas' lemma (Exercise 8) to deduce the existence of (u, v) with

$$u \geqslant 0, \qquad v \geqslant 0, \qquad uA + v = c, \qquad \langle u, b \rangle \geqslant \langle c, \bar{x} \rangle.$$

Prove that u is feasible in the dual and optimal. In particular, both programs have the same value.

Exercise 10. Values of antisymmetric games (Brown and von Neumann [34])

(1) The aim is to prove that any real $I \times I$ matrix B which is *antisymmetric* $(B = -{}^t B)$ has a value. Let $X = \Delta(I)$ be the simplex of mixed strategies.

   (a) Show that it is equivalent to find $x \in X(B) = \{x \in X \text{ with } Bx \leqslant 0\}$.
   (b) Let $K^i(x) = [e^i Bx]^+$, $\bar{K}(x) = \sum_i K^i(x)$ and consider the dynamical system on X defined by:

$$\dot{x}^i(t) = K^i(x(t)) - x^i(t)\bar{K}(x(t)) \quad (*)$$

   Show that the set of rest points of $(*)$ is $X(B)$. Let $V(x) = \sum_i K^i(x)^2$. Check that $V^{-1}(0) = X(B)$ and show that $V(x(t))$ is strictly decreasing on the complement of $X(B)$. Conclude.

(2) We now deduce from 1(a) that any matrix A has a value. Show that one can assume $A_{ij} > 0$ for all (i, j).

   (a) Following (Gale et al. [73]), let us introduce the antisymmetric matrix B of size $(I + J + 1) \times (I + J + 1)$ defined by

$$B = \begin{pmatrix} 0 & A & -1 \\ -{}^t A & 0 & 1 \\ 1 & -1 & 0 \end{pmatrix}.$$

   Construct from an optimal strategy z for player 1 in game B optimal strategies for both players in game A.
   (b) An alternative proof is to consider the $(I \times J) \times (I \times J)$ matrix C defined by

$$C_{ij;i'j'} = A_{ij'} - A_{i'j}.$$

   Hence each player plays in game C both as player 1 and 2 in the initial game A. Show that an optimal strategy in game C allows us to construct optimal strategies for both players in game A.

Exercise 11. Strong complementarity

(1) Use Farkas' lemma to prove, with the notations of Exercise 9:
   If $\mathcal{P}^*$ and $\mathcal{P}$ are feasible, then there exist optimal solutions (x, u) satisfying

$$A_i x - b_i > 0 \Longleftrightarrow u_i = 0 \; ; \quad c_j - u A_j > 0 \Longleftrightarrow x_j = 0.$$

(2) Deduce (c) of Proposition 3.1.

Exercise 12. Fictitious play (Robinson [177])

Let A be a real $I \times J$ matrix of size $m \times n$. We describe an algorithm that will produce optimal strategies and in particular will imply the existence of a value.

Let $\underline{v} = \max_{x \in \Delta(I)} \min_{y \in \Delta(J)} x A y$ and $\overline{v} = \min_{y \in \Delta(J)} \max_{x \in \Delta(I)} x A y$.

Given $\alpha \in \mathbb{R}^n$ (resp. $\beta \in \mathbb{R}^m$) define

$$J(\alpha) = \{j; \alpha^j = \min_{k \in J} \alpha^k\}, \quad I(\beta) = \{i; \beta^i = \max_{k \in I} \beta^k\}.$$

An *admissible* sequence of vectors $\{(\alpha(t), \beta(t)) \in \mathbb{R}^n \times \mathbb{R}^m, t \in \mathbb{N}\}$ satisfies:

(i) $\min_{k \in J} \alpha^k(0) = \max_{k \in I} \beta^k(0)$

and there exists an $i \in I(\beta(t))$ and $j \in J(\alpha(t))$ such that

(ii) $\alpha(t+1) = \alpha(t) + e^i A, \qquad \beta(t+1) = \beta(t) + A e^j.$

(In particular, $\alpha(t) - \alpha(0) = t x(t) A$ for some $x(t) \in \Delta(I)$.)

(1) Show that for any $\varepsilon > 0$ and t large enough

$$\min_{j \in J} \alpha^j(t)/t \leqslant \underline{v} + \varepsilon, \qquad \max_{i \in I} \beta^i(t)/t \geqslant \overline{v} - \varepsilon,$$

and, considering the dual game $^t A$ and using (i), that for any $s \in \mathbb{N}$

(iii) $\min_{i \in I} \beta^i(s) - \max_{j \in J} \alpha^j(s) \leqslant 0.$

(2) One says that i (resp. j) is *useful* in the interval $[t_1, t_2] \subset \mathbb{N}$ if there exists a t with $t_1 \leqslant t \leqslant t_2$ and $i \in I(\beta(t))$ (resp. $j \in J(\alpha(t)))$. Show that if all $j \in J$ are useful in $[s, s+t]$, then

(iv) $\max_{j \in J} \alpha^j(s+t) - \min_{j \in J} \alpha^j(s+t) \leqslant 2t\|A\|.$

Define $\mu(t) = \max_{i \in I} \beta^i(t) - \min_{j \in J} \alpha^j(t)$. Deduce from (iii) and (iv) that if all $i \in I, j \in J$ are useful in $[s, s+t]$, then

(v) $\mu(s+t) \leqslant 4t\|A\|.$

(3) We want to prove:

(§) For any matrix A and any $\varepsilon > 0$, there exists an s such that $\mu(t) \leqslant \varepsilon$, for all $t \geqslant s$ and all admissible sequences.

The proof is by induction on the size of A (clear for $m = n = 1$). Assume that (§) holds for all (strict) sub-matrices A' of A.

(a) Show that there exists a $t^* \in \mathbb{N}$ such that $\mu'(t) \leqslant \frac{\varepsilon}{2} t$ for all $t \geqslant t^*$, all A', and all associated admissible sequences α', β', where μ' is defined for A' as μ is for A.

(b) Show that if $i \in I$ is not useful in $[s, s+t^*]$, then we have

(vi) $\mu(s+t^*) \leqslant \mu(s) + \frac{\varepsilon}{2} t^*.$

(c) Let $t = qt^* + r$ be the Euclidean division of t by t^* and let $p \leqslant q$ be the largest integer such that all i, j are useful in $[(p - 1)t^* + r, pt^* + r]$ (and $p = 0$ if this is never the case). Prove using (vi) that

$$\mu(t) \leqslant \mu(pt^* + r) + \frac{\varepsilon}{2}(q - p)t^*,$$

then using (v) that

$$\mu(t) \leqslant 4\|A\|t^* + \frac{\varepsilon}{2}t.$$

(d) Deduce property (§).

(4) Show that the matrix A has a value and that any accumulation point of $\{x(t)\}$ (resp. $\{y(t)\}$) is an optimal strategy of player 1 (resp. 2).

2.9 Comments

The notion of a game goes back to the 18th century but the introduction of mixed strategies is a 20th century development, due to Borel [31]. The initial proof of the minmax theorem by von Neumann [224] uses a fixed point argument. The first "modern" proof relying on a separation theorem is due to Ville [222] and is employed in *Theory of Games and Economic Behavior* [225, p. 154]. The same tool was used to establish the duality theorem for linear programming in the same period.

The proofs of the minmax theorem can be classified as follows:

- The finite case and recurrence argument: Fourier's elimination algorithm, Loomis' proof (Chap. 2, Exercise 1), Fictitious play and Robinson's proof (Chap. 2, Exercise 12).
- Elementary separation (in $\mathbb{R}^n$), see Proposition 3.3.3.
- The algorithmic approach (unilateral): Blackwell approachability (Chap. 2, Exercise 4), no-regret procedures (Chap. 7).
- The dynamic approach (global): Fictitious play or Best response dynamics, Brown–von Neumann (Chap. 2, Exercise 10), replicator dynamics (Chap. 5).

Many interesting properties of the value operator and of the set of optimal strategies can be found in the *Contributions to the Theory of Games*, see [47, 48, 113, 114]. They play a crucial rôle in the "operator approach" to zero-sum repeated games.

The zero-sum paradigm is also very important when evaluating the strength of player i facing an unknown environment defined by $-i$, see elimination of strategies (Chap. 4) and the individually rational level (Chap. 8).

Chapter 3
Zero-Sum Games: The General Case

3.1 Introduction

This chapter deals with the general case of a zero-sum game, where the strategy sets of the players may be infinite. A zero-sum game will always be a triple (S, T, g) where S and T are non-empty sets and g is a mapping from $S \times T$ to $\mathbb{R}$. By definition, the game has a value if $\sup_{s \in S} \inf_{t \in T} g(s, t) = \inf_{t \in T} \sup_{s \in S} g(s, t)$, and minmax theorems, such as von Neumann minmax theorem in the previous Chap. 2, refer to results providing sufficient conditions on the triple (S, T, g) for the existence of a value. Recall that if the value exists, a strategy s in S achieving the supremum in $\sup_{s \in S} \inf_{t \in T} g(s, t)$ is called an optimal strategy of player 1 in the game. Similarly, $t \in T$ achieving the infimum in $\inf_{t \in T} \sup_{s \in S} g(s, t)$ is called optimal for player 2.

We prove here various minmax theorems. We start with Sion's theorem for convex compact action sets and payoff functions which are quasi-concave upper semi-continuous in the first variable and quasi-convex lower semi-continuous in the second variable. Then we prove the standard minmax theorem in mixed strategies, extending von Neumann's theorem to compact Hausdorff action sets and measurable bounded payoff functions which are u.s.c. in the first variable and l.s.c. in the second variable. Finally, we consider the value operator (strategy sets are fixed and the payoff function varies) and its directional derivatives, and introduce the derived game.

3.2 Minmax Theorems in Pure Strategies

The following result is known as the *intersection lemma* (see [22, p. 172]), and will be useful later.

Lemma 3.2.1 *Let $C_1, \ldots, C_n$ be non-empty convex compact subsets of a Euclidean space. Assume that the union $\bigcup_{i=1}^{n} C_i$ is convex and that for each $j = 1, \ldots, n$, the intersection $\bigcap_{i \neq j} C_i$ is non-empty. Then the full intersection $\bigcap_{i=1}^{n} C_i$ is also non-empty.*

© Springer Nature Switzerland AG 2019

R. Laraki et al., *Mathematical Foundations of Game Theory*, Universitext,

https://doi.org/10.1007/978-3-030-26646-2_3

Proof By induction on n. For $n = 2$, consider C_1 and C_2 satisfying the assumptions of the lemma, and assume that $C_1 \cap C_2 = \varnothing$. By the Hahn–Banach theorem these two sets can be strictly separated by some hyperplane H. $C_1 \cup C_2$ is convex and contains a point in each of the two half-spaces with boundary H, so $C_1 \cup C_2$ must intersect H, which is a contradiction.

Consider now $n \geqslant 3$ and assume by contradiction that the lemma holds for $n - 1$ but not for n. Let $C_1, \ldots, C_n$ satisfy the hypotheses of the lemma and have empty intersection. Then C_n and $\bigcap_{i=1}^{n-1} C_i$ (denoted by D_n in the sequel) are non-empty, disjoint, convex and compact. Again these two sets can be strictly separated by some hyperplane H.

Define the convex compact sets $\tilde{C}_i = C_i \cap H$ for $i = 1, \ldots, n - 1$, and $\tilde{C} = (\bigcup_{i=1}^n C_i) \cap H$. Since $C_n \cap H = \varnothing = D_n \cap H$, we have $\bigcup_{i=1}^{n-1} \tilde{C}_i = \tilde{C}$ and $\bigcap_{i=1}^{n-1} \tilde{C}_i = \varnothing$. By the induction hypothesis applied to $\tilde{C}_1, \ldots, \tilde{C}_{n-1}$, there exists a j in $\{1, \ldots, n - 1\}$ such that $\bigcap_{i \neq j,n} \tilde{C}_i = \varnothing$. Introduce $K = \bigcap_{i \neq j,n} C_i$, then $D_n \subset K$ and $C_n \cap K \neq \varnothing$. As K is convex and intersects sets separated by the hyperplane H, we have $K \cap H \neq \varnothing$. But $K \cap H = \bigcap_{i \neq j,n} \tilde{C}_i = \varnothing$, which is a contradiction. $\qquad\square$

Remark 3.2.2 Since the proof only uses the Hahn–Banach strict separation theorem, the result holds in every space where this theorem applies, in particular in every locally convex Hausdorff topological vector space.

Before stating Sion's theorem, we recall a few definitions.

Definition 3.2.3 If E is a convex subset of a vector space, a map $f : E \to \mathbb{R}$ is *quasi-concave* if for each λ in $\mathbb{R}$, the upper section $\{x \in E, f(x) \geqslant \lambda\}$ is convex. f is *quasi-convex* if $-f$ is quasi-concave.

If E is a topological space, a map $f : E \to \mathbb{R}$ is *upper semi-continuous* (u.s.c.) if for each λ in $\mathbb{R}$, the upper section $\{x \in E, f(x) \geqslant \lambda\}$ is closed. f is *lower semi-continuous* (l.s.c.) if $-f$ is u.c.s.

Clearly, if E is compact and f is u.s.c., then f has a maximum.

In the following, the strategy sets S and T are subsets of Hausdorff topological real vector spaces.

Theorem 3.2.4 ([191]) *Let $G = (S, T, g)$ be a zero-sum game satisfying:*

  (i) S and T are convex;
  (ii) S or T is compact;
  (iii) for each t in T, $g(\cdot, t)$ is quasi-concave u.s.c. in s, and for each s in S, $g(s, \cdot)$ is quasi-convex l.s.c. in t.

  Then G has a value:

$$\sup_{s \in S} \inf_{t \in T} g(s, t) = \inf_{t \in T} \sup_{s \in S} g(s, t).$$

Moreover, if S (resp. T) is compact, the above suprema (resp. infima) are achieved, and the corresponding player has an optimal strategy.

Proof Assume S compact. By upper semi-continuity, $\sup_{s \in S} \inf_{t \in T} g(s, t) = \max_{s \in S} \inf_{t \in T} g(s, t)$. So the existence of an optimal strategy for player 1 will follow from the existence of the value.

Suppose by contradiction that G has no value. Then there exists a real number v such that

$$\sup_{s \in S} \inf_{t \in T} g(s, t) < v < \inf_{t \in T} \sup_{s \in S} g(s, t).$$

(1) We first reduce the problem to the case where S and T are polytopes. Define for each t in T the set $S_t = \{s \in S, g(s, t) < v\}$. The family $(S_t)_{t \in T}$ is an open covering of the compact set S, so there exists a finite subset T_0 of T such that $S = \bigcup_{t \in T_0} S_t$. Define $T' = \text{co}(T_0)$ to be the convex hull of the finite set T_0. T' is compact (each finite-dimensional Hausdorff topological vector space has a unique topology), and $\max_{s \in S} \inf_{t \in T'} g(s, t) < v < \inf_{t \in T'} \sup_{s \in S} g(s, t)$.

Proceed similarly with the strategy space of player 1: the family $(T'_s = \{t \in T', g(s, t) > v\})_{s \in S}$ being an open covering of T', there exists a finite subset S_0 of S such that

$$\forall s \in \text{co}(S_0), \exists t \in T_0, \quad g(s, t) < v,$$

$$\forall t \in \text{co}(T_0), \exists s \in S_0, \quad g(s, t) > v.$$

(2) Moreover, we can assume without loss of generality that (S_0, T_0) is a minimal pair for inclusion satisfying this property: If necessary drop elements from S_0 and/or T_0.

(3) For each s in S_0, define now $A_s = \{t \in \text{co}(T_0), g(s, t) \leqslant v\}$. A_s is a non-empty convex compact subset of $\text{co}(T_0)$. Note that $\bigcap_{s \in S_0} A_s = \varnothing$ and by minimality of S_0, $\bigcap_{s \in S_0 \setminus \{s_0\}} A_s \neq \varnothing$ for each s_0 in S_0. By the intersection lemma, the union $\bigcup_{s \in S_0} A_s$ is not convex. Consequently there exists a t_0 in $\text{co}(T_0) \setminus \bigcup_{s \in S_0} A_s$. For each s in S_0, $g(s, t_0) > v$. By quasi-concavity of $g(\cdot, t_0)$, the inequality $g(s, t_0) > v$ also holds for each s in $\text{co}(S_0)$.

Similarly, we show the existence of some s_0 in $\text{co}(S_0)$ such that $g(s_0, t) < v$ for each t in $\text{co}(T_0)$. By considering $g(s_0, t_0)$, we find a contradiction with G having no value. $\square$

We can weaken the topological conditions by strengthening the convexity hypothesis on $g(s, \cdot)$. In the next result no topology on T is needed.

Proposition 3.2.5 *Let $G = (S, T, g)$ be a zero-sum game such that:*

(i) S is convex and compact;
(ii) T is convex;
(iii) for each t in T, $g(\cdot, t)$ is quasi-concave u.s.c., and for each s in S, $g(s, \cdot)$ is convex.

Then G has a value: $\sup_{s \in S} \inf_{t \in T} g(s, t) = \inf_{t \in T} \sup_{s \in S} g(s, t)$, and player 1 has an optimal strategy.

Proof Suppose there exists a real number v satisfying

$$\sup_{s \in S} \inf_{t \in T} g(s, t) < v < \inf_{t \in T} \sup_{s \in S} g(s, t).$$

As in the proof of Sion's theorem, there exists a finite set $T_0 = \{t_1, \ldots, t_J\} \subset T$ such that $\forall s \in S, \exists t \in T_0, g(s, t) < v$.

Let us endow the affine space generated by T_0 with any norm, and denote by $\text{int}(\text{co}(T_0))$ the relative interior of $\text{co}(T_0)$. For each s in S, $g(s, \cdot)$ is convex, hence continuous on $\text{int}(\text{co}(T_0))$ (a convex function defined on an open subset of a Euclidean space is continuous on it, see e.g. Berge [22, Theorem 7, p. 203]). Fix t_0 in $\text{int}(\text{co}(T_0))$ and define for each $n \geq 1$ and $j \in \{1, \ldots, J\}$: $t_j^n = \frac{1}{n}t_0 + (1 - \frac{1}{n})t_j$ and $S_{t_j}^n = \{s \in S, g(s, t_j^n) < v\}$. For all j and n, the convexity of $g(s, \cdot)$ gives $g(s, t_j^n) \leq \frac{1}{n}g(s, t_0) + (1 - \frac{1}{n})g(s, t_j)$. So the sets $S_{t_j}^n$ form an open covering of S, and there exists a finite subset T_1 of $\text{int}(\text{co}(T_0))$ such that

$$\forall s \in S, \exists t \in T_1, \quad g(s, t) < v,$$

$$\forall t \in \text{co}(T_1), \exists s \in S, \quad g(s, t) > v.$$

$g(\cdot, t)$ is u.s.c. for each t, and $g(s, \cdot)$ is continuous on $\text{co}(T_1)$ for each s in S. Hence

$$\max_{s \in S} \min_{t \in \text{co}(T_1)} g(s, t) < \min_{t \in \text{co}(T_1)} \max_{s \in S} g(s, t).$$

This contradicts Sion's theorem, so G has a value.

Since the map $(s \mapsto \inf_{t \in T} g(s, t))$ is u.s.c., player 1 has an optimal strategy (but not necessarily player 2). $\qquad\square$

3.3   Minmax Theorems in Mixed Strategies

Recall (Sect. 1.3) that given a game (S, T, g), whenever possible the payoff function g is linearly extended to probability distributions, i.e. $g(\sigma, \tau) = \mathbb{E}_{\sigma \otimes \tau}(g)$ for σ and τ probabilities on S and T, respectively. The probability σ, resp. τ, is called a mixed strategy of player 1, resp. player 2. We will also sometimes call elements of S and T pure strategies of player 1 and 2, and saying that the game (S, T, g) has a value in pure strategies simply means $\sup_{s \in S} \inf_{t \in T} g(s, t) = \inf_{t \in T} \sup_{s \in S} g(s, t)$.

We consider here games without any convexity assumption on the strategy spaces. We will convexify a set X either by considering the set $\Delta_f(X)$ of probabilities with finite support over X ($\Delta_f(X)$ being regarded as the convex hull of X) or, if X has a topological structure, by considering the set $\Delta(X)$ of regular probabilities over X (i.e. Borel probability measures μ such that for each Borel subset A of X, $\mu(A) = \sup\{\mu(F), F \subset A, F \text{ compact}\} = \inf\{\mu(G), G \supset A, G \text{ open}\}$; if X is metric compact, any Borel probability on X is regular).

Recall that in this case $\Delta(X)$ is endowed with the weak* topology (the weakest topology such that $\hat{\phi}: \mu \mapsto \int_X \phi \, d\mu$ is continuous for each real continuous function ϕ on X). Then if X is Hausdorff compact, $\Delta(X)$ is also compact and if ϕ is u.s.c. on X, $\hat{\phi}: \mu \mapsto \int_X \phi \, d\mu$ is u.s.c. on $\Delta(X)$ (see e.g. Kelley and Namioka [107]). In the case when X metric, μ_n converges weakly* to μ if $\int_X \phi \, d\mu_n \to \int_X \phi \, d\mu$ for each real continuous function ϕ on X and $\Delta(X)$ is metric (see e.g. Billingsley [25]; Parthasarathy [159]).

Let us start with a minmax theorem based on Proposition 3.2.5.

Proposition 3.3.1 *Consider a zero-sum game (S, T, g) satisfying:*

(i) S *is a compact Hausdorff topological space;*
(ii) *for each t in T, $g(\cdot, t)$ is u.s.c.*

Then the game $(\Delta(S), \Delta_f(T), g)$ has a value and player 1 has an optimal strategy.

Proof Recall that if S is compact and $g(\cdot, t)$ u.s.c., then $\Delta(S)$ (endowed with the weak* topology) is compact and $(\sigma \mapsto g(\sigma, t) = \int_S g(s, t)\sigma(ds))$ is u.s.c. Moreover, $g(\sigma, \tau)$ is well defined on $\Delta(S) \times \Delta_f(T)$ and is bilinear. Then the assumptions of Proposition 3.2.5 are satisfied. □

The following result is the standard minmax theorem in mixed strategies. We assume the payoff function of the game measurable and bounded, so that we can apply Fubini's theorem and define the mixed extension of the game.

Theorem 3.3.2 *Let $G = (S, T, g)$ be a zero-sum game such that:*

(i) S *and T are compact Hausdorff topological spaces;*
(ii) *for each t in T, $g(\cdot, t)$ is u.s.c., and for each s in S, $g(s, \cdot)$ is l.s.c.;*
(iii) g *is bounded and measurable with respect to the product Borel σ-algebra $\mathcal{B}_S \otimes \mathcal{B}_T$.*

Then the mixed extension $(\Delta(S), \Delta(T), g)$ of G has a value. Each player has a mixed optimal strategy, and for each $\varepsilon > 0$ each player has an ε-optimal strategy with finite support.

Proof One can apply Proposition 3.3.1 to the games $G^+ = (\Delta(S), \Delta_f(T), g)$ and $G^- = (\Delta_f(S), \Delta(T), g)$ and obtain the values v^+ and v^-, respectively. Clearly $v^- \leq v^+$.

Let σ (resp. τ) be an optimal strategy of player 1 in the game G^+ (resp. player 2 in G^-). We have

$$\int_S g(s, t)\sigma(ds) \geq v^+, \quad \forall t \in T,$$

$$\int_T g(s, t)\tau(dt) \leq v^-, \quad \forall s \in S.$$

So by Fubini's theorem

$$v^+ \leq \int \int_{S \times T} g(s, t)\sigma(ds)\tau(dt) \leq v^-$$

and the result follows. □

Minmax theorems can also be obtained using separation theorems in Euclidean spaces.

Proposition 3.3.3 *Assume:*

(i) *S is a measurable space and X is a non-empty convex set of probabilities over S;*
(ii) *T is a finite non-empty set;*
(iii) *$g : S \times T \longrightarrow \mathbb{R}$ is measurable and bounded.*

Then the game $(X, \Delta(T), g)$ has a value, and player 2 has an optimal strategy.

Proof Define $\underline{v} = \sup_X \inf_T g(x, t)$ and $D = \{a \in \mathbb{R}^T : \exists x \in X, g(x, t) = \int_X g(s, t)x(ds) = a^t, \forall t \in T\}$. The set D is convex and disjoint from the convex set $C = \{a \in \mathbb{R}^T; a t \geqslant \underline{v} + \varepsilon, \forall t \in T\}$, where $\varepsilon > 0$ is arbitrary.

Using a standard separation theorem in the Euclidean space $\mathbb{R}^T$, we obtain the existence of a $b \neq 0$ in $\mathbb{R}^T$ such that

$$\langle b, d \rangle \leqslant \langle b, c \rangle \qquad \forall c \in C, \forall d \in D.$$

C is positively comprehensive, so $b \geqslant 0$, and dividing b by $\sum_{t \in T} b_t$ gives the existence of $y \in \Delta(T)$ satisfying

$$g(x, y) \leqslant \underline{v} + \varepsilon \quad \forall x \in X.$$

So $\overline{v} := \inf_{\Delta(T)} \sup_X g(x, y) \leqslant \underline{v} + \varepsilon$. Hence $\overline{v} = \underline{v}$ and the value of $(X, \Delta(T), g)$ exists. The existence of an optimal strategy for player 2 follows by the compactness of $\Delta(T)$.   □

3.4   The Value Operator and the Derived Game

Here we fix the strategy sets S and T and let the payoff functions vary. Consider a set $\mathcal{F}$ of functions from $S \times T$ to $\mathbb{R}$ such that:

(a) $\mathcal{F}$ is a convex cone (i.e., $\mathcal{F}$ is stable under addition and multiplication by a positive scalar, and $0 \in \mathcal{F}$) containing the constant functions; and
(b) for each f in $\mathcal{F}$ the game (S, T, f) has a value, which we denote by $\mathrm{val}_{S \times T}(f) = \sup_{s \in S} \inf_{t \in T} f(s, t) = \inf_{t \in T} \sup_{s \in S} f(s, t)$, or simply by $\mathrm{val}(f)$.

Clearly, the val operator:

(1) is *monotonic*: $f \leqslant g \Rightarrow \mathrm{val}(f) \leqslant \mathrm{val}(g)$; and
(2) *translates constants*: $\forall t \in \mathbb{R}, \mathrm{val}(f + t) = \mathrm{val}(f) + t$.

An easy corollary follows, where one uses the sup norm $\|f\| = \sup_{S \times T} |f(s, t)|$.

Proposition 3.4.1 *The* val *operator is non-expansive:*

$$|\mathrm{val}(f) - \mathrm{val}(g)| \leqslant \|f - g\|.$$

Proof $f \leqslant g + \|f - g\|$ implies, by (1), $\mathrm{val}(f) \leqslant \mathrm{val}(g + \|f - g\|)$ and the last term is $\mathrm{val}(g) + \|f - g\|$, by 2). $\qquad\square$

The following proposition is due to Mills [142] in the case where S and T are simplices and $\mathcal{F}$ is the set of bilinear functions over $S \times T$. It was extended later by Rosenberg and Sorin [180].

Proposition 3.4.2 *Consider two compact sets S and T, and two real-valued functions f and g defined on $S \times T$. Assume that for each $\alpha \geqslant 0$, the functions g and $f + \alpha g$ are u.s.c. in s and l.s.c. in t, and that the zero-sum game $(S, T, f + \alpha g)$ has a value $\mathrm{val}_{S \times T}(f + \alpha g)$. Then*

$$\lim_{\alpha \to 0^+} \frac{1}{\alpha} \left[\mathrm{val}_{S \times T}(f + \alpha g) - \mathrm{val}_{S \times T}(f) \right] \quad exists$$

and is equal to

$$\mathrm{val}_{S(f) \times T(f)}(g),$$

where $S(f)$ and $T(f)$ are the respective sets of optimal strategies of player 1 and 2 in the game (S, T, f).

Proof The compactness and semi-continuity hypotheses imply the existence of $s_\alpha \in S(f + \alpha g)$ and $t \in T(f)$. We have

$$\alpha g(s_\alpha, t) = [f + \alpha g](s_\alpha, t) - f(s_\alpha, t) \geqslant \mathrm{val}_{S \times T}(f + \alpha g) - \mathrm{val}_{S \times T}(f).$$

So

$$\inf_{t \in T(f)} g(s_\alpha, t) \geqslant \frac{1}{\alpha} \left[\mathrm{val}_{S \times T}(f + \alpha g) - \mathrm{val}_{S \times T}(f) \right]$$

and

$$\limsup_{\alpha \to 0^+} \inf_{T(f)} g(s_\alpha, t) \geqslant \limsup_{\alpha \to 0^+} \frac{1}{\alpha} \left[\mathrm{val}_{S \times T}(f + \alpha g) - \mathrm{val}_{S \times T}(f) \right].$$

Let $(\alpha_k)_k$ be a vanishing sequence achieving $\limsup_{\alpha \to 0^+} \inf_{T(f)} g(s_\alpha, t)$ and s^* be a limit point of $(s_{\alpha_k})_k$. Since g is u.c.s. in s, we have

$$\inf_{T(f)} g(s, t) \geqslant \limsup_{\alpha \to 0^+} \inf_{T(f)} g(s_\alpha, t).$$

Moreover, $s^* \in S(f)$ (S is compact, $f + \alpha g$ is u.s.c. in s and g is l.s.c. in t), so

$$\sup_{S(f)} \inf_{T(f)} g(s, t) \geqslant \limsup_{\alpha \to 0^+} \frac{1}{\alpha} \left[\mathrm{val}_{S \times T}(f + \alpha g) - \mathrm{val}_{S \times T}(f) \right]$$

and the result follows from the dual inequality.   □

The game $(S(f), T(f), g)$ is called the *derived game* of f along the game g.

3.5 Exercises

Exercise 1. Duels
We follow Dresher [46]. Two players start a duel at a distance $d(0) > 0$ from each other. They move towards each other, and the distance between the two players at time $t \in [0, 1]$, if both are still alive, is given by $d(t)$. Assume that d is a strictly decreasing function from $d(0)$ to $d(1) = 0$ (at time $t = 1$, if both are still alive, the players are at the same point).

Each player has a gun with one or several bullets, and chooses when he will shoot at his opponent. For each player i, denote by $p_i(t)$ the probability that this player will kill his opponent if he shoots at time t, and assume that with probability $1 - p_i(t)$ his opponent is not even touched by the bullet. Assume also that p_1 and p_2 are strictly increasing continuous functions with $p_1(0) = p_2(0) = 0$ and $p_1(1) = p_2(1) = 1$: shooting immediately always fails, and shooting someone at distance zero always succeeds.

p_1 and p_2 are known by both players. The payoff of a player is: $+1$ if he is alone to survive the duel, -1 if he dies and his opponent survives, and 0 otherwise.

(1) We assume that each player has a single bullet and that the duel is noisy. So if a player is the first one to shoot and misses his opponent, the other player will know it and eventually win the duel with probability one by shooting point blank. A pure strategy of a player can thus be represented here by a number x in $[0, 1]$, giving the time when this player will shoot* at his opponent (meaning: if no bullet has been used before).

Show that the game has a value in pure strategies and that the optimal pure strategy of each player is to shoot at time t_0 such that $p_1(t_0) + p_2(t_0) = 1$.

(2) The duel is still noisy but now player 1 has m bullets and player 2 has n bullets (and this is known to both players). For simplicity, we assume that $p_1(t) = p_2(t) = t$ for each t.

Show by induction on $n + m$ that the game has a value (possibly in random strategies) equal to $\frac{m-n}{m+n}$ and that it is optimal for the player with $\max\{m, n\}$ bullets to shoot the first bullet at time $t_0 = \frac{1}{m+n}$.

(3) Each player has a single bullet, but the guns are now silent: a player does not know whether his opponent has already shot at him. A pure strategy of a player is still represented by a number x in $[0, 1]$, giving the time when this player will

shoot at his opponent if he is still alive at that time. Again, we assume that $p_1(t) = p_2(t) = t$ for each t.

(a) Show that the game has no value in pure strategies.
(b) Suppose player 1 shoots according to a mixed strategy with support $[\alpha, 1]$ and density f. Show that there exist a differentiable f and $\alpha > 0$ such that whatever the strategy of player 2, the payoff of player 1 is non-negative. Conclude.

(4) Both players have a single bullet, but here the gun of player 1 is silent whereas player 2 has a noisy gun. Assume that $p_1(t) = p_2(t) = t$ for each t. Let us prove that the game has a value v in mixed strategies, where $v = 1 - 2a$ with $a = \sqrt{6} - 2$.

(a) Show that the mixed strategy:

$$f(x) = \begin{cases} 0 & \text{if } 0 \leqslant x < a \\ \dfrac{\sqrt{2a}}{\left(x^2 + 2x - 1\right)^{3/2}} & \text{if } a \leqslant x \leqslant 1 \end{cases}$$

guarantees $1 - 2a$ to player 1.
(b) Show that player 2 guarantees the same amount by playing the mixed strategy with the following cumulative distribution function:

$$G(y) = \frac{2}{2+a} \int_0^y f(x)dx + \frac{a}{2+a} I_1(y),$$

where $I_1(y)$ is the c.d.f. associated to the Dirac measure at 1. Here player 2 uses, with probability $\frac{2}{2+a}$, the same strategy as player 1, and shoots at time 1 with the remaining probability.

(5) Is it possible to compute the value by induction for silent duels with several bullets?

Exercise 2. A counter-example (Sion [191])
Define $S = T = [0, 1]$ and $f : S \times T \to \{0, -1\}$ by:

$$f(s, t) = \begin{cases} -1 & \text{if} \quad t = 0 \text{ and } s < \frac{1}{2} \\ -1 & \text{if} \quad t = 1 \text{ and } s \geqslant \frac{1}{2} \\ 0 & \text{otherwise.} \end{cases}$$

(1) Show that the game has no value in pure strategies, and that the hypotheses of Sion's theorem are satisfied everywhere except at $t = 1$.
(2) Does this game have a value in mixed strategies?

Exercise 3. A monotone family
We consider a family of zero-sum games $G_n = (S, T, f_n)$ such that:

– (f_n) is a weakly decreasing sequence of uniformly bounded functions from $S \times T$ to $\mathbb{R}$, u.s.c. in the first variable;
– for each n, G_n has a value v_n;
– S is compact.

(1) Define $f = \inf_n f_n$. Show that the game $G = (S, T, f)$ has a value $v = \inf_n v_n$, and that player 1 has an optimal strategy in this game.
(2) Compare the value v of G and $\lim v_n$ in the following two examples of one-player games:

    (i) $S = [0, +\infty[$, $f_n(s) = \mathbf{1}_{\{s \geqslant n\}}$;
    (ii) $S = [0, 1]$, f_n is continuous and piecewise linear (affine on $[0, 1/n]$, on $[1/n, 2/n]$ and on $[2/n, 1]$): $f_n(0) = f_n(\frac{2}{n}) = f_n(1) = 0$ and $f_n(\frac{1}{n}) = 1$.

Exercise 4. Blackwell approachability
Let $A = (a_{i,j})$ be an $I \times J$-matrix with entries in $\mathbb{R}^K$: $a_{i,j} \in \mathbb{R}^k$ is the vector payoff, or outcome, if player 1 plays i and player 2 plays j.

    Given $s \in \Delta(I)$, we denote by sA the subset of $\mathbb{R}^K$ of feasible expected vector payoffs when player 1 plays the mixed action s:

$$
\begin{aligned}
sA &= \{z \in \mathbb{R}^k : \exists t \in \Delta(J) \quad \text{s.t.} \quad z = sAt\} \\
&= \Big\{ \textstyle\sum_{i \in I, j \in J} s_i A_{i,j} t_j, t \in \Delta(J) \Big\} \\
&= \text{co} \Big\{ \textstyle\sum_{i \in I} s_i A_{i,j}, j \in J \Big\}.
\end{aligned}
$$

Let C be a closed convex subset of $\mathbb{R}^K$. For each x in $\mathbb{R}^K$, endowed with the Euclidean norm, denote by $\Pi_C(x)$ the closest point to x in C, i.e. the projection of x on C.

    Assume that C is a **B**-set for player 1, i.e. satisfies:

$$
\forall x \notin C, \exists s \in \Delta(I) \quad \text{s.t.} \quad \forall z \in sA : \quad \langle z - \Pi_C(x), x - \Pi_C(x) \rangle \leqslant 0.
$$

Geometrically, the affine hyperplane containing $\Pi_C(x)$ and orthogonal to $[x, \Pi_C(x)]$ separates x from sA.

    The game is played in discrete time for infinitely many stages: at each stage $n = 1, 2, \ldots$, after having observed the past history h_{n-1} of actions chosen from stage 1 to stage $n - 1$, i.e. $h_{n-1} = (i_1, j_1, \ldots, i_{n-1}, j_{n-1}) \in \mathcal{H}_{n-1}$, (with $\mathcal{H}_n = (I \times J)^n$ for each n and $\mathcal{H}_0 = \{\varnothing\}$), player 1 chooses $s_n(h_{n-1}) \in \Delta(I)$ and simultaneously player 2 chooses $t_n(h_{n-1}) \in \Delta(J)$. Then a pair $(i_n, j_n) \in I \times J$ is selected according to the product probability $s_n(h_{n-1}) \otimes t_n(h_{n-1})$. The play then goes to stage $n + 1$ with the history $h_n = (i_1, j_1, \ldots, i_n, j_n) \in \mathcal{H}_n$.

    Consequently, a strategy σ of player 1 in the repeated game takes the form of a sequence $\sigma = (s_1, \ldots, s_n, \ldots)$ with $s_n : \mathcal{H}_{n-1} \to \Delta(I)$ for each n, and a strategy τ of player 2 is an element $\tau = (t_1, \ldots, t_n, \ldots)$ with $t_n : \mathcal{H}_{n-1} \to \Delta(J)$. A pair (σ, τ) naturally defines a probability distribution $\mathbb{P}_{\sigma,\tau}$ over the set of plays $\mathcal{H}_\infty = (I \times J)^\infty$, endowed with the product σ-algebra, and we denote by $\mathbb{E}_{\sigma,\tau}$ the associated expectation.

Every play $h = (i_1, j_1, \ldots, i_n, j_n, \ldots)$ of the game induces a sequence of vector payoffs $x(h) = (x_1 = a_{i_1,j_1}, \ldots, x_n = a_{i_n,j_n}, \ldots)$ with values in $\mathbb{R}^K$. We denote by $\bar{x}_n$ the Cesàro average up to stage n:

$$\bar{x}_n(h) = \frac{1}{n} \sum_{k=1}^{n} a_{i_k,j_k} = \frac{1}{n} \sum_{k=1}^{n} x_k.$$

Blackwell [27] constructed a strategy σ of player 1 which generates a play $h = (i_1, j_1, \ldots, i_n, j_n, \ldots)$ such that $\bar{x}_n(h)$ converges to C, whatever the strategy τ of player 2:

$$d_n = \|\bar{x}_n - \Pi_C(\bar{x}_n)\| \xrightarrow[n \to \infty]{} 0, \mathbf{P}_{\sigma,\tau}\text{-a.s.}$$

Blackwell's strategy σ is defined inductively as follows: At stage $n + 1$, play $s_{n+1} \in \Delta(I)$ such that, for each $t \in \Delta(J)$,

$$\langle s_{n+1}At - \Pi_C(\bar{x}_n), \bar{x}_n - \Pi_C(\bar{x}_n) \rangle \leq 0.$$

This definition uses the fact that C is a **B**-set. Notice that if $\bar{x}_n \in C$, any s_{n+1} will do and player 1 can play arbitrarily.

(1) Show that

$$\mathbf{E}_{\sigma,\tau}\left[d_{n+1}^2 \,|h_n\right] \leq \frac{1}{(n+1)^2} \mathbf{E}_{\sigma,\tau}\left[\|x_{n+1} - \Pi_C(\bar{x}_n)\|^2 \,|h_n\right] + \left(\frac{n}{n+1}\right)^2 d_n^2.$$

(2) Prove that $\mathbf{E}_{\sigma,\tau}\left[\|x_{n+1} - \Pi_C(\bar{x}_n)\|^2 \,|h_n\right] \leq 4\|A\|_\infty^2$, where $\|A\|_\infty = \max_{i,j,k} \left\|A_{i,j}^k\right\|$.

(3) Deduce that:

$$\mathbf{E}_{\sigma,\tau}[d_n] \leq \frac{2\|A\|_\infty}{\sqrt{n}}.$$

In particular, the convergence is uniform in τ.

(4) Define $e_n = d_n^2 + \sum_{k=n+1}^{\infty} \frac{4\|A\|_\infty^2}{k^2}$ for each n. Show that $\{e_n\}$ is a positive supermartingale whose expectation converges to 0. Conclude that

$$\mathbf{P}_{\sigma,\tau}[d_n \to 0] = 1.$$

Exercise 5. **sup inf** and **inf sup**

Let f be a function from $S \times T$ to $\mathbb{R}$, where S and T are arbitrary non-empty sets. Denote by B the set of mappings from S into T. Show that

$$\sup_{s \in S} \inf_{t \in T} f(s, t) = \inf_{\beta \in B} \sup_{s \in S} f(s, \beta(s)).$$

Exercise 6. On mixed extension

Consider the game $G = (S, T, f)$, where $S = T = (0, 1]$ and

$$
f(s, t) = \begin{cases} 0 & \text{if } s = t \\ -\dfrac{1}{s^2} & \text{if } s > t \\ \dfrac{1}{t^2} & \text{if } s < t. \end{cases}
$$

(1) Show that for each $t \in T$, $\int_S f(s, t)\, ds = 1$.
(2) Show that: $\sup_{\sigma \in \Delta(S)} \inf_{t \in T} f(\sigma, t) > \inf_{\tau \in \Delta(T)} \sup_{s \in S} f(s, \tau)$, where $f(\sigma, t) = \int_S f(s, t)\, d\sigma(s)$ and $f(s, \tau) = \int_T f(s, t)\, d\tau(t)$.
(3) Recall that $\sup_\sigma \inf_\tau f \leqslant \inf_\tau \sup_\sigma f$. What should one think of the mixed extension of G?

Exercise 7. Sion and Wolfe [192]

Consider $S = T = [0, 1]$ endowed with the Borel σ-algebra, and f defined on $S \times T$ by:

$$
f(s, t) = \begin{cases} -1 & \text{if } s < t < s + 1/2, \\ 0 & \text{if } t = s \text{ or } t = s + 1/2, \\ 1 & \text{otherwise.} \end{cases}
$$

Consider the mixed extension G where player 1 chooses σ in $\Delta(S)$, player 2 chooses τ in $\Delta(T)$, and the payoff of player 1 is $f(\sigma, \tau) = \int_{S \times T} f(s, t)\, d\sigma(s)\, d\tau(t)$.

(1) Show that $\sup_{\sigma \in \Delta(S)} \inf_{t \in T} f(\sigma, t) = 1/3$.
(2) Prove that G has no value.

3.6   Comments

Note that the main result of this chapter, Sion's theorem, relies in fact on a separation theorem in finite dimensions.

There is a huge literature on minmax theorems in topological spaces (Kneser, Wald, etc), see for instance [137, Chap. 1]. In particular we see there how to approximate or to regularize a triple (S, T, g) not satisfying Sion's requirements.

When there is no linear structure one can use the Concave-like and the Convex-like properties on g to purify mixed strategies [54] (which corresponds to the inverse procedure: deducing the existence of pure optimal strategies from the results in Sect. 3.3).

The properties of the value operator and the derived game play an important rôle in the operator approach to zero-sum repeated games.

Exercise 1, question 2, is a first example of the "recursive structure" that occurs in repeated games.

Chapter 4
N-Player Games: Rationality and Equilibria

4.1 Introduction

The previous chapters dealt with two-player zero-sum games. Now we study strategic interactions with two or more players where, in addition to conflict, there is now room for cooperation, but also problems of coordination and free riding.

Nash equilibrium is probably the central solution concept of game theory, with applications in economics, biology, computer science, political science, and elsewhere. It is a strategy profile such that no player has an incentive to unilaterally deviate. Nash's famous theorem says that any finite n-person game admits a (mixed) equilibrium. As in Nash's original proofs, we will deduce equilibrium existence from the Brouwer and Kakutani fixed point theorems, whose proofs will also be given.

This chapter then goes beyond the classical finite framework, and studies games with arbitrarily compact strategy spaces and possibly discontinuous payoff functions. We provide tight geometric and topological conditions for the existence of rationalizable strategy profiles, Nash equilibria and approximate equilibria. These concepts have deep connections with each other. Actually, any Nash equilibrium is rationalizable and is an approximate equilibrium.

We also investigate some properties of Nash equilibria such as geometry in finite games, characterization via variational inequalities, conditions for uniqueness, and invariance with respect to some payoff transformations.

To make a link with the previous chapters, when a game is zero-sum, the existence of a Nash equilibrium is equivalent to the existence of a value and of optimal strategies, and the existence of an approximate equilibrium is analogous to the existence of a value.

4.2 Notation and Terminology

In this chapter, $G = (I, (S^i)_{i \in I}, (g^i)_{i \in I})$ will denote a game in strategic (or normal) form, where I is the non-empty finite set of players, S^i is the non-empty set of pure strategies of player $i \in I$ and $g^i : S = \prod_{j \in I} S^j \to \mathbb{R}$ represents its payoff function.

© Springer Nature Switzerland AG 2019
R. Laraki et al., *Mathematical Foundations of Game Theory*, Universitext,
https://doi.org/10.1007/978-3-030-26646-2_4

The game is *finite* if each strategy set S^i is finite. Using Reny's [174] terminology, the game is called *compact* if each strategy set S^i is a compact subset of a topological space and each payoff function g^i is bounded. Finally, the game G is called *continuous* if each S^i is a topological space and each payoff function g^i is continuous with respect to the product topology on S.

For a coalition $J \subset I$, introduce $S^J = \prod_{j \in J} S^j$. The coalition $I \setminus \{i\}$ will be denoted, as usual, by $-i$, so that $S^{I \setminus \{i\}} = S^{-i} = \prod_{j \neq i} S^j$ and $S^I = S$.

A *correlated strategy* of players in J is an element $\theta^{[J]} \in \Delta(S^J)$ and a *mixed strategy profile* of players in J is an element $\sigma^J \in \prod_{j \in J} \Delta(S^j)$, where $\Delta(S^J)$ is the set of regular probability measures on the Borel subsets of the topological space S^J.

4.3   Best Response Domination in Finite Games

We assume in this section that the game $G = (I, (S^i)_{i \in I}, (g^i)_{i \in I})$ is finite.

In Chap. 1, the best response correspondence was defined from/to the set of pure strategy profiles. It can be extended to a correspondence defined on the set of correlated strategies with values in the set of mixed strategy profiles. First, observe that the payoff function g^i can be linearly extended to $\Delta(S)$ as usual: for $\theta \in \Delta(S)$, $g^i(\theta) := \sum_{s \in S} \theta(s) g^i(s)$. In particular, if $m^i \in \Delta(S^i)$ and $\theta^{[-i]} \in \Delta(S^{-i})$ then

$$
g^i(m^i, \theta^{[-i]}) = \sum_{s=(s^i,s^{-i}) \in S} m^i(s^i) \times \theta^{[-i]}(s^{-i}) \times g^i(s^i, s^{-i})
$$

$$
= \sum_{s^i \in S^i} m^i(s^i) g^i(s^i, \theta^{[-i]}).
$$

Definition 4.3.1 BR^i is the *general best response correspondence* of player i. It associates to every $\theta^{[-i]} \in \Delta(S^{-i})$ the set $\{m^i \in \Delta(S^i) : g^i(m^i, \theta^{[-i]}) \geq g^i(\sigma^i, \theta^{[-i]})$ for all $\sigma^i \in \Delta(S^i)\}$.

By linearity of the map $m^i \to g^i(m^i, \theta^{[-i]})$, $m^i \in \mathrm{BR}^i(\theta^{[-i]})$ is equivalent to $g^i(m^i, \theta^{[-i]}) \geq g^i(s^i, \theta^{[-i]})$ for all $s^i \in S^i$. Consequently, $\mathrm{BR}^i(\theta^{[-i]})$ is a simplex (a face of $\Delta(S^i)$) whose extreme points are the pure best responses of player i against $\theta^{[-i]}$.

Definition 4.3.2 A mixed strategy $m^i \in \Delta(S^i)$ is *strictly dominated* if there is a $\sigma^i \in \Delta(S^i)$ such that for all $t^{-i} \in S^{-i}$, $g^i(\sigma^i, t^{-i}) > g^i(m^i, t^{-i})$.

A mixed strategy $m^i \in \Delta(S^i)$ is *never a best response against a mixed strategy profile* of $-i$ if there is no $\tau^{-i} \in \Pi_{j \neq i} \Delta(S^j)$ such that $m^i \in \mathrm{BR}^i(\tau^{-i})$.

A mixed strategy $m^i \in \Delta(S^i)$ is *never a best response against a correlated strategy* of $-i$ if there is no $\theta^{[-i]} \in \Delta(S^{-i})$ such that $m^i \in \mathrm{BR}^i(\theta^{[-i]})$.

Remark 4.3.3 In two-player games, the last two definitions are the same.

   As soon as there are at least three players, a strategy may be a best response to a correlated strategy but not to a mixed strategy, as the following example shows.

	L	R			L	R			L	R			L	R
T	8	0		T	4	0		T	0	0		T	3	3
B	0	0		B	0	4		B	0	8		B	3	3

$$M^1 \qquad\qquad M^2 \qquad\qquad M^3 \qquad\qquad M^4$$

   Here player 1 chooses between T and B, player 2 between L and R, and player 3 in the set $\{M^i, i = 1, \dots, 4\}$. The payoffs above are those of player 3. One can check that M^2 is a best response against the correlated strategy $\frac{1}{2}(T, L) + \frac{1}{2}(B, R)$ (meaning that players 1 and 2 play (T, L) with probability $\frac{1}{2}$ and (B, R) with probability $\frac{1}{2}$). But M^2 is never a best response against a mixed profile of players 1 and 2 (there is no x and y in $[0, 1]$ such that M^2 is a best response against player 1 playing $xT + (1 - x)B$ and, independently, player 2 playing $yL + (1 - y)R$).

Proposition 4.3.4 *A mixed strategy $m^i \in \Delta(S^i)$ of player i is strictly dominated if and only if it is never a best response against a correlated strategy of $-i$.*

Proof Let m^i be strictly dominated by $\sigma^i \in \Delta(S^i)$. By linearity of g^i in $\theta^{[-i]}$, one has $g^i(\sigma^i, \theta^{[-i]}) > g^i(m^i, \theta^{[-i]})$ for all $\theta^{[-i]} \in \Delta(S^{-i})$: m^i cannot be a best response against a correlated strategy of $-i$.

   Conversely, suppose that m^i is never a best response against a correlated strategy. Consider the mixed extension of the finite two-player zero-sum matrix game H where player i, the maximizer, has the strategy set S^i and the minimizer (the team $-i$) has the strategy set S^{-i}. The payoff function of player i in H is $h^i(t^i, t^{-i}) = g^i(t^i, t^{-i}) - g^i(m^i, t^{-i})$. The hypothesis on m^i implies that the value of H is strictly positive. Consequently, any optimal mixed strategy σ^i of player i in H strictly dominates m^i.   □

Definition 4.3.5 A mixed strategy $m^i \in \Delta(S^i)$ is *weakly dominated* if there is a $\sigma^i \in \Delta(S^i)$ such that for all $t^{-i} \in S^{-i}$, $g^i(\sigma^i, t^{-i}) \geqslant g^i(m^i, t^{-i})$ and there is at least one t^{-i} for which the inequality is strict.

   A mixed strategy $m^i \in \Delta(S^i)$ is never a best response against a completely correlated strategy of $-i$ if there is no $\theta^{[-i]} \in \text{int}(\Delta(S^{-i}))$ (i.e. $\theta^{[-i]}(s^{-i}) > 0, \forall s^{-i} \in S^{-i}$) s.t. $m^i \in \text{BR}^i(\theta^{[-i]})$.

Proposition 4.3.6 *A mixed strategy $m^i \in \Delta(S^i)$ of player i is weakly dominated if and only if it is never a best response against a completely correlated strategy of players $-i$.*

Proof Let m^i be weakly dominated by $\sigma^i \in \Delta(S^i)$. Linearity of g^i in $\theta^{[-i]}$ implies $g^i(\sigma^i, \theta^{[-i]}) > g^i(m^i, \theta^{[-i]})$ for all $\theta^{[-i]} \in \text{int}(\Delta(S^{-i}))$. Thus, m^i cannot be a best response against a completely correlated strategy of $-i$.

Suppose m^i is not a best response against a completely correlated strategy. Consider the mixed extension of the finite zero-sum matrix game H as defined in the previous proof. Since player i can guarantee 0 in H by playing m^i, the value of H is at least 0. If the value is strictly positive, then any optimal strategy of player i in H strictly dominates m^i in G.

Suppose the value of H equals zero. This implies in particular that m^i is an optimal strategy. Because m^i is never a best response to a completely correlated strategy, no optimal strategy of $-i$ has full support, and there is a $t^{-i} \in S^{-i}$ which does not belong to the support of an optimal strategy of $-i$. Since in finite zero-sum games a pure strategy belongs to the support of some optimal strategy if and only if it is a best response against all optimal strategies of the opponent (see Proposition 2.4.1 d), there is a σ^i optimal for player i in H against which t^{-i} is not a best response. Thus, for all s^{-i}, $h^i(\sigma^i, s^{-i}) \geqslant 0 = h^i(m^i, s^{-i})$ and $h^i(\sigma^i, t^{-i}) > 0 = h^i(m^i, t^{-i})$. We conclude that σ^i weakly dominates m^i in G. $\square$

4.4 Rationalizability in Compact Continuous Games

In this section, the game $G = (I, (S^i)_{i \in I}, (g^i)_{i \in I})$ is assumed compact and continuous (for example, the mixed extension of a finite game, as studied in the previous section).

A strategy which is never a best response against a strategy profile will not be played by a player who maximizes against some strategy of the opponents: one says that it cannot be *justified*. In particular, a strictly dominated strategy is not justified (for example, if the compact continuous game G is the mixed extension of a finite game G_0, an unjustified strategy in G corresponds to a mixed strategy in G_0 which is never a best response to a mixed strategy profile in G_0). Once all players eliminate non-best response strategies, new strategies become unjustified, and so on.

This defines inductively a process of eliminations of unjustified strategies. At step 1 of this process, define

$$S^i(1) = \mathrm{BR}^i(S^{-i}) = \{s^i;\, \exists s^{-i} \in S^{-i},\ s^i \text{ is a best response to } s^{-i}\}, \forall i \in I.$$

Then, inductively, at step $k+1$, let $S^i(k+1) = \mathrm{BR}^i(S^{-i}(k))$. Thus, at step $k+1$, strategies that are unjustified with respect to $S(k)$ are eliminated. This leads to a decreasing sequence whose limit is $S^i_\infty := \bigcap_k S^i(k)$. Let $S_\infty = \prod_{i \in I} S^i_\infty$. Elements $t \in S_\infty$ are called *rationalizable*.

Proposition 4.4.1 ([23, 161]) *Let G be a compact continuous game. Then S_∞ is a non-empty compact fixed point of* BR. *It is the largest set $L \subset S$ satisfying the property*

$$L \subset \mathrm{BR}(L).$$

Proof The continuity and compactness of G implies the upper-semicontinuity of the BR correspondence and by induction, the non-emptiness and compactness of $S(k)$ for every k. Since moreover $S(k)$ is decreasing, the intersection $S_\infty = \bigcap_k S(k)$ is non-empty and compact. Since $S_\infty \subset S(k+1) = \mathrm{BR}(S(k))$,

$$S_\infty \subset \lim_{k \to \infty} \mathrm{BR}(S(k)) \subset \mathrm{BR}(S_\infty)$$

(by upper-semicontinuity of BR). Finally, since $S_\infty \subset S(k)$ implies that $\mathrm{BR}(T_\infty) \subset \mathrm{BR}(S(k)) = S(k+1)$ for all k, we obtain $\mathrm{BR}(S_\infty) \subset S_\infty$.

Now, let $L \subset S$ be such that $L \subset \mathrm{BR}(L)$. Because $L \subset S = S(1)$ and $L \subset \mathrm{BR}(L)$ we deduce that $L \subset \mathrm{BR}(S(1)) = S(2)$. Inductively, $L \subset S(k)$ and $L \subset \mathrm{BR}(L)$ implies that $L \subset \mathrm{BR}(S(k)) = S(k+1)$. Since this holds for every k, we obtain that $L \subset S_\infty$. This implies that S_∞ is the largest fixed set of BR. □

Remarks 4.4.2

- If G is the mixed extension of a finite game, Proposition 4.3.4 provides a link between elimination of strategies that are never a best response against a correlated strategy and iterated elimination of strictly dominated strategies.
- If G is the mixed extension of a finite game, the process stops in finite time.
- If all players know their payoff function and play some best response, then the play must be in $S(1)$. Moreover, if all players know the game (i.e. all parameters S^i and g^i) and that players use a best response, the play must be in $S(2)$, and if all players know that all players know the game and play a best response, then the play must be in $S(3)$. Pursuing this reasoning inductively leads to a play in S_∞ and the process is related to *common knowledge of rationality*.
- Some learning and evolutionary game processes (such as the replicator dynamics, see Chap. 5), which assume a low level of knowledge, lead remarkably to a play in S_∞. In rationalizability, there is no time variable and players, by a recursive reasoning, conclude simultaneously that the play should be in S_∞. In learning and evolutionary game theory, the repetition of the interaction induces the players to recursively stop playing the strategies that perform badly in the past, hence as times goes on, only strategies in S_∞ survive.

Definition 4.4.3 A game is *solvable* if the set S_∞ of rationalizable outcomes is reduced to a singleton.

Guess $\frac{2}{3}$ of the average is a game where each player i in I sends a guess a^i in $[0, 100]$, and the winner is the one with the guess closest to $\frac{2}{3}$ of the average $v = \frac{\sum_{i \in I} a^i}{N}$ of all guesses. Clearly, guessing any number that lies above $\frac{2}{3}100$ is never a best response. These can be eliminated. Once these strategies are eliminated for every player, bids above $\frac{4}{9}100$ are never best responses and can be eliminated. This process will continue until reaching a unique rationalizable outcome $S_\infty = \{0\}$.

4.5 Nash and ε-Equilibria: Definition

A Nash equilibrium of a strategic game is a strategy profile such that no player has
a unilateral profitable deviation. This is the minimal stability criterion one may
ask a profile to satisfy. There are many fields where Nash equilibrium is applied:
economics, political science, biology, philosophy, language, and computer science,
among others. Here, this notion is considered as satisfying a mathematical condition,
and our objective is to study its existence and some of its properties.

Definition 4.5.1 A *Nash equilibrium* of a game $G = (I, (S^i)_{i \in I}, (g^i)_{i \in I})$ is a strategy profile $s = (s^i)_{i \in I} \in S$ such that $g^i(t^i, s^{-i}) \leqslant g^i(s^i, s^{-i})$ for all i in I and t^i in
S^i.

There are different formulations of the Nash equilibrium condition. One way of
describing it is that $s^i \in \mathrm{BR}^i(s^{-i})$ for every $i \in I$, which says that every player is best
replying to the other players' strategy profile. This is the most common formulation.
Proving existence with this formulation is equivalent to showing that the best response
correspondence has a fixed point. Under some regularity assumptions this correspondence will satisfy Kakutani or Brouwer's conditions and so will admit a fixed point (as
will be seen in the next two sections). A more subtle formulation defines a dominance
binary relation between strategy profiles, namely, say that strategy profile t dominates
s and write $t \succ s$ if there is an $i \in I$ such that $g^i(t^i, s^{-i}) > g^i(s^i, s^{-i})$. Equivalently,
s is not dominated by t and write $s \succeq t$ if $g^i(s^i, s^{-i}) \geqslant g^i(t^i, s^{-i})$, $\forall i \in I$. Hence,
a Nash equilibrium is a strategy profile which is un-dominated (or maximal). Such
a formulation is interesting when strategy sets are infinite-dimensional or payoff
functions are discontinuous (see Sects. 4.7 and 4.8).

   In some games, Nash equilibria may not exist (or may be hard to compute). In
that case, one may be interested in ε-equilibria (see Chap. 1).

Definition 4.5.2 For any $\varepsilon \geqslant 0$, an *ε-equilibrium* is a strategy profile $s \in S$ such that
for every player i, $s^i \in \mathrm{BR}^i_\varepsilon(s^{-i})$, that is,

$$g^i(t^i, s^{-i}) \leqslant g^i(s) + \varepsilon, \quad \forall t^i \in S^i, \quad \forall i \in I.$$

   A Nash equilibrium corresponds to $\varepsilon = 0$. Hence, the pure best response correspondence, as defined in Chap. 1, BR from $S \rightrightarrows S$, associates to each $s \in S$ the
subset $\prod_{i \in I} \mathrm{BR}^i_0(s^{-i})$ of S.

Definition 4.5.3 A strategy profile $s \in S$ is a *strict equilibrium* if $\{s\} = \mathrm{BR}(s)$.

   When $\mathrm{BR}(s)$ is reduced to a singleton $\{f(s)\}$ for every $s \in S$, a Nash equilibrium
is always strict and is a fixed point of f.

Corollary 4.5.4 *If s is a Nash equilibrium, then it is rationalizable.*

Proof We have $\{s\} \subset \mathrm{BR}(\{s\})$ and S_∞ is the largest subset L of S satisfying $L \subset$
$\mathrm{BR}(L)$. $\square$

Corollary 4.5.5 *If the game is solvable, it has a unique Nash equilibrium.*

In Sect. 4.7.3 we will give other conditions for the uniqueness of a Nash equilibrium.

Remark 4.5.6 In contrast with zero-sum games, a non-zero-sum game may have several equilibrium payoffs, and moreover, equilibrium strategies are not interchangeable. For example, in a common interest game where a player gets l if all players choose the same location $l \in \{1, \ldots, L\}$ and gets zero otherwise, any location choice l by all players is a Nash equilibrium, equilibria have different payoffs, but all players have the same preference among the equilibria. On the other hand, in a bargaining game, two players should agree on dividing 1 unit among them, otherwise they get 0. Any division $(x, 1 - x)$ with $x \in [0, 1]$ is a Nash equilibrium, but players have opposite preferences among the equilibria.

4.6   Nash Equilibrium in Finite Games

Recall (see Chap. 1) that $\tilde{G} = (I, (\Delta(S^i))_{i \in I}, (\tilde{g}^i)_{i \in I})$, the mixed extension of a finite game $G = (I, (S^i)_{i \in I}, (g^i)_{i \in I})$, is the compact multilinear continuous game where the strategy set of player i is $\Delta(S^i)$ (the set of probability distributions over S^i) and the payoff function of $\tilde{G}$ is the multilinear extension of the payoff function in G:

$$\tilde{g}^i(\sigma) = \sum_{s=(s^1,\ldots,s^N) \in S} \left(\prod_{j \in I} \sigma^j(s^j) \right) g^i(s).$$

When there is no confusion, $\tilde{g}^i$ will also be denoted by g^i.

Definition 4.6.1 A *mixed equilibrium* of G is a Nash equilibrium of $\tilde{G}$.

Theorem 4.6.2 ([152]) *Every finite game G has a mixed equilibrium.*

The proof uses the finiteness of the game and the linearity of the payoff functions with respect to each player's strategy variable to reduce the problem to the existence of a fixed point of a continuous mapping from $\prod_{i \in I} \Delta(S^i)$ to itself. By Brouwer's Theorem 4.11.4 (see the last section of this chapter) a mixed equilibrium exists.

Proof Let f be the Nash map, defined from $\Delta := \prod_{i \in I} \Delta(S^i)$ to Δ as follows:

$$f(\sigma)^i(s^i) = \frac{\sigma^i(s^i) + (g^i(s^i, \sigma^{-i}) - g^i(\sigma))^+}{1 + \Sigma_{t^i \in S^i}(g^i(t^i, \sigma^{-i}) - g^i(\sigma))^+}$$

with $a^+ = \max(a, 0)$. The function f is well defined and takes its values in Δ: $f(\sigma)^i(s^i) \geqslant 0$ and $\sum_{s^i \in S^i} f(\sigma)^i(s^i) = 1$, for all $i \in I$. Since f is continuous and Δ

is convex and compact, Brouwer's Theorem 4.11.4 implies the existence of $\sigma \in \Delta$ such that $f(\sigma) = \sigma$. Such σ is a Nash equilibrium, as proved next.

Fix a player i. If $\Sigma_{t^i \in S^i}(g^i(t^i, \sigma^{-i}) - g^i(\sigma))^+ = 0$ then $g^i(\sigma^i, \sigma^{-i}) \geqslant \max_{t^i \in S^i} g^i(t^i, \sigma^{-i})$: player i plays a best response to the other players' strategies. Otherwise, $\Sigma_{t^i \in S^i}(g^i(t^i, \sigma^{-i}) - g^i(\sigma))^+ > 0$. But since there exists an s^i with $\sigma^i(s^i) > 0$ and $g^i(s^i, \sigma^{-i}) \leqslant g^i(\sigma)$, we obtain

$$\sigma^i(s^i) = \frac{\sigma^i(s^i)}{1 + \Sigma_{t^i \in S^i}(g^i(t^i, \sigma^{-i}) - g^i(\sigma))^+}$$

and consequently $\sigma^i(s^i) = 0$, a contradiction. Thus, σ is a mixed equilibrium.

Conversely, any mixed equilibrium is a fixed point of f, since all quantities $(g^i(t^i, \sigma^{-i}) - g^i(\sigma))^+$, $i \in I$, $t^i \in S^i$, are equal to zero.   □

A game G has symmetry ϕ if:

(1) ϕ is a permutation over the set N of players, and, if $j = \phi(i)$, ϕ induces a bijection from S^i to S^j, also denoted ϕ; and
(2) for all $i \in I$ and $s \in S$, $g^{\phi(i)}(\phi(s)) = g^i(s)$.

Such a permutation ϕ naturally induces a mapping on Δ such that if $j = \phi(i)$ and $\sigma \in \Delta$ then $\phi(\sigma)^j(\phi(s^i)) = \sigma^i(s^i)$.

Theorem 4.6.3 ([152]) *A finite game G with symmetry ϕ admits a mixed equilibrium σ with the same symmetry (i.e. $\sigma = \phi(\sigma)$).*

Proof Let $X \subset \Delta$ be the subset of mixed strategy profiles with symmetry ϕ:

$$X = \{\sigma \in \Delta \text{ such that } \sigma = \phi(\sigma)\}.$$

Then, X is non-empty because the strategy profile where all players play uniformly is in X. Moreover, X is closed and convex. The Nash mapping f constructed in the previous proof obviously respects symmetry ($f(X) \subseteq X$). Brouwer's Theorem 4.11.4 ensures the existence of a fixed point in X.   □

4.7 Nash Equilibrium in Continuous Games

In this section, $G = (I, (S^i)_{i \in I}, (g^i)_{i \in I})$ is a compact continuous game. We will provide topological and geometrical conditions for the existence of pure Nash equilibria, from which we deduce the existence of mixed equilibria in all compact continuous games. Finally, under more assumptions, we characterize Nash equilibria via variational inequalities and provide a monotonicity condition that implies uniqueness.

4.7.1 Existence of Equilibria in Pure Strategies

To prove the existence of Nash equilibria, we need to assume some topological and geometrical conditions, in particular we will assume—in this section—that each strategy set S^i is a convex subset of a topological vector space (TVS).

Recall that a real-valued function f on a convex set X is quasi-concave if the level sets $\{x : f(x) \geqslant \alpha\}$ are convex for all reals α.

Definition 4.7.1 A game is *quasi-concave* if for all $i \in I$ the map $s^i \mapsto g^i(s^i, s^{-i})$ is quasi-concave for all s^{-i} in S^{-i} and all $i \in I$.

Theorem 4.7.2 *If a game G is compact, continuous and quasi-concave, then its set of Nash equilibria is a non-empty and compact subset of $\prod_{i \in I} S^i$.*

Hence, the topological conditions are the compactness of the strategy sets and the continuity of the payoff functions. The geometrical assumptions are the convexity of the strategy sets and the quasi-concavity of the payoff functions w.r.t. each player's decision variable.

Proof First, we use the maximal element formulation of the Nash equilibrium problem in addition to the geometric and topological conditions to reduce the problem to finite-dimensional convex strategy sets. Second, we prove that the best-response correspondence of the restricted game satisfies the assumptions of the Kakutani fixed point Theorem 4.11.5 (proved in the last section of this chapter) to conclude the existence of a fixed point, hence of a Nash equilibrium.

Define $A(t) = \{s \in S$ such that $g^i(s^i, s^{-i}) \geqslant g^i(t^i, s^{-i}), \forall i \in I\}$. Hence $A(t)$ is the set of strategy profiles not dominated by t (no component of t is a profitable deviation). Consequently, s is a Nash equilibrium if and only if $s \in \bigcap_{t \in S} A(t)$. The continuity and compactness of the game imply that $A(t)$ is a compact subset of S. Thus, the set of Nash equilibria is compact and existence holds if $\bigcap_{t \in S} A(t)$ is non-empty. Now using the finite intersection property it is sufficient to prove that for every finite set (of potential deviations) $t_0, t_1, \ldots, t_k$ in S, $\bigcap_{t \in \{t_0, t_1, \ldots, t_k\}} A(t)$ is non-empty.

Let $\Delta_k = \{\alpha = (\alpha_0, \ldots, \alpha_k) \in \mathbb{R}^{k+1} : \alpha_l \geqslant 0$ and $\sum_{l=0}^{k} \alpha_l = 1\}$ be the k-dimensional simplex of the Euclidean space $\mathbb{R}^{k+1}$ and let ϕ^i be the map from Δ_k to S^i defined as follows. For $\alpha = (\alpha_0, \ldots, \alpha_k) \in \Delta_k$, $\phi^i(\alpha) = \sum_{l=0}^{k} \alpha_l t_l^i$. Because S^i is a convex subset of a topological vector space, ϕ^i is continuous and has values in $\mathrm{co}\{t_0^i, \ldots, t_k^i\} \subset S^i$, where co stands for the convex hull.

Now we can define the game $\widehat{G}$ where the strategy set of each player $i \in I$ is $\Delta_k^i = \Delta_k$ and its payoff is $f^i(\alpha^1, \ldots, \alpha^I) = g^i(\phi^1(\alpha^1), \ldots, \phi^I(\alpha^I))$. Also, because $s^i \mapsto g^i(s^i, s^{-i})$ is quasi-concave, $\alpha^i \mapsto g^i(\alpha^i, \alpha^{-i})$ is quasi-concave. Thus, the game $\widehat{G}$ is compact, continuous, quasi-concave and the strategy set of each player $i \in I$ is the k-simplex (a convex compact subset of a normed space).

Let $\mathrm{BR}_{\widehat{G}}$ (from $\prod_{i \in I} \Delta_k^i$ to itself) be the best-response correspondence of the game $\widehat{G}$. The quasi-concavity of $\widehat{G}$ implies that for all α, $\mathrm{BR}_{\widehat{G}}(\alpha)$ is convex. The continuity and compactness of $\widehat{G}$ imply that for all $\alpha \in \prod_{i \in I} \Delta_k^i$, $\mathrm{BR}_{\widehat{G}}(\alpha)$ is non-empty and that the graph of $\mathrm{BR}_{\widehat{G}}$ is closed. Applying Kakutani's Theorem 4.11.5, we deduce that the set of fixed points of $\mathrm{BR}_{\widehat{G}}$ is compact and non-empty. That is, there exists an $\alpha \in \mathrm{BR}_{\widehat{G}}(\alpha)$. Consequently, if we define $s^i = \phi^i(\alpha^i)$ for every $i \in I$, then $s \in \bigcap_{t \in \{t_0, t_1, \ldots, t_k\}} A(t)$.   □

Theorem 4.7.2 was first proved by Glicksberg [79] and Fan [53] when the strategy sets are subsets of a locally convex and Hausdorff topological vector space. It has been extended by Reny [174] to any topological vector space and to a large class of discontinuous functions (see Corollary 4.8.6 below).

4.7.2   Existence of Equilibria in Mixed Strategies

When a strategic game G satisfies the right topological conditions (compactness of strategy sets and continuity of payoff functions), one can show that its mixed extension $\tilde{G} = (I, (\Delta(S^i))_{i \in I}, (\tilde{g}^i)_{i \in I})$ satisfies the right topological and geometrical conditions, leading to the existence of a mixed equilibrium, thanks to Theorem 4.7.2.

Theorem 4.7.3 *If a game G is compact and continuous, then its set of mixed equilibria is a non-empty compact subset of $\prod_{i \in I} \Delta(S^i)$.*

Proof Recall (Sect. 3.2) that whenever S^i is compact, $\Delta(S^i)$ is also compact with respect to the weak* topology (see [107, 132]). Also, a Stone–Weierstrass type argument (see [132, 137]) implies that each payoff function $g^i(s)$ can be ε-approximated by a linear combination of separable functions $g_\varepsilon^i(s) = \sum_k \alpha_k \prod_{i \in I} g_\varepsilon^{i,k}(s^i)$ where each $g_\varepsilon^{i,k}(s_i)$ is a continuous function on the compact topological space S^i.

This implies that payoff functions in $\tilde{G}$ are continuous. Finally, since a payoff function in $\tilde{G}$ is multilinear, $\tilde{G}$ is a quasi-concave game. Thus, in the weak* topology, the game $\tilde{G}$ is compact, continuous and quasi-concave and the strategy sets are convex subsets of a topological vector space. Theorem 4.7.2 applied to $\tilde{G}$ implies that the set of mixed equilibria of G is non-empty and compact.   □

Remark 4.7.4 When S^i is a metric compact space the theorem can be proved by an approximation argument. Let S_n^i be an increasing sequence of finite subsets of S^i such that $\bigcup_n S_n^i$ is dense in S^i. By Nash's Theorem 4.6.2, for every n, the finite game $G_n = (I, (S_n^i)_{i \in I}, (g^i)_{i \in I})$ admits a Nash equilibrium σ_n. Since $\Delta(S^i)$ is compact and metric in the weak* topology [50], there is a subsequence $\phi(n)$ of the integers such that $\sigma_{\phi(n)}^i$ converges to $\sigma^i \in \Delta(S^i)$ for every $i \in I$. Since $g^i(\sigma_n) \geqslant g^i(s_n^i, \sigma_n^{-i})$ for every $n, i \in I$ and $s_n^i \in S_n^i$, by continuity, $g^i(\sigma) \geqslant g^i(s^i, \sigma^{-i})$ for every $i \in I$ and $s^i \in S^i$.

4.7.3   Characterization and Uniqueness of Nash Equilibria

In this subsection, we assume that the strategy sets are convex subsets of a Hilbert space and we call such a game a *Hilbert game*.

When the payoff functions are smooth, one can derive some necessary first-order conditions for a strategy profile to be a Nash equilibrium. Under additional geometrical assumptions, those conditions are also sufficient.

Definition 4.7.5 The game is *smooth* if $s^i \mapsto g^i(s^i, s^{-i})$ is C^1 for all s^{-i} in S^{-i} and all $i \in I$. It is *concave* if $s^i \mapsto g^i(s^i, s^{-i})$ is concave for all s^{-i} in S^{-i} and all $i \in I$.

In a smooth game, let $\nabla_i g^i(s)$ denote the gradient of $g^i(s^i, s^{-i})$ with respect to s^i.

Theorem 4.7.6 *Let* $G = (I, \{S^i\}_{i \in I}, \{g^i\}_{i \in I})$ *be a smooth game. Then:*

(1) If s is a Nash equilibrium then

$$\langle \vec{\nabla} g(s), s - t \rangle \geqslant 0, \quad \forall t \in S,$$

where $\langle \vec{\nabla} g(s), s - t \rangle := \sum_{i \in I} \langle \nabla_i g^i(s), s^i - t^i \rangle$.
(2) If the game is concave, the condition in (1) is sufficient for s to be a Nash equilibrium.

This is a direct consequence of the following classical lemma.

Lemma 4.7.7 *Let* $f : X \to \mathbb{R}$ *be a* C^1 *function over a convex subset X of a Hilbert space.*

(1) If x locally maximizes f over X then, $\forall y \in X$, $\langle \nabla f(x), x - y \rangle \geqslant 0$.
(2) If f is concave, the above condition is sufficient for x to be a global maximum over X.

Above, $\nabla f(x)$ is the gradient of f with respect to x.

Proof If x is a local maximum of f on the convex set X, then for every y in X and every $t \in]0, 1]$, one has $\frac{f(x+t(y-x))-f(x)}{t} \leqslant 0$. Letting t goes to zero implies $\langle \nabla f(x), x - y \rangle \geqslant 0$ for every $y \in X$.

When f is concave, for every x and y in X, $f(y) \leqslant f(x) + \langle \nabla f(x), y - x \rangle$. Thus, if for some $x \in X$, $\langle \nabla f(x), x - y \rangle \geqslant 0$ for every $y \in X$, then $f(y) \leqslant f(x)$ for every $y \in X$. Consequently, x is a global maximum. □

Example 4.7.8 In the 19th century Cournot [38] introduced the strategic equilibrium concept to study his duopoly competition model. Each firm $i \in \{1, 2\}$ chooses to produce a quantity $q^i \in [0, a]$. The cost function of firm i is linear in quantity: $C^i(q^i) = cq^i$. The market price p is also assumed to be linear on the total production: $p = \max\{a - (q^1 + q^2); 0\}$, where $a > c > 0$. When firm i chooses its "strategy" q^i its payoff is

$$g^i(q^1, q^2) = p \times q^i - C_i(q^i) = \max\{a - (q^1 + q^2); 0\}q^i - cq^i.$$

This is a compact continuous game where the payoff function of each player i is quasi-concave in q^i. Thus, by Theorem 4.7.2, there is at least one Cournot–Nash equilibrium. It is easy to see that this equilibrium cannot be at the boundary ($q^i = 0 \Rightarrow q^j =$

$\frac{a-c}{2} \Rightarrow q^i = \frac{a-c}{4}$). Because any best response is in $[0, \frac{a-c}{2}]$ and in this domain the payoff function is concave and smooth, the first-order condition is necessary and sufficient (by Theorem 4.7.6). Consequently $\mathrm{BR}^i(q^j) = \frac{a-c-q^j}{2}$, implying $q^1 = \frac{a-c-q^2}{2}$ and $q^2 = \frac{a-c-q^1}{2}$. Thus the unique Cournot–Nash equilibrium is $q^1 = q^2 = \frac{a-c}{3}$. One can show that this game is dominance solvable. As calculated above, any best response should be in $S_1 = [a_1, b_1] = [0, \frac{a-c}{2}]$, a second iteration implies that any best response to S_1 must be in $S_2 = [a_2, b_2] = [\frac{a-c-b_1}{2}, \frac{a-c-a_1}{2}] = [\frac{a-c}{4}, \frac{a-c}{2}]$, etc. Repeated elimination of non-best response strategies leads to $S_\infty = \{\frac{a-c}{3}\}$.

The fact that the Cournot duopoly game above has a unique equilibrium may also be deduced from its monotonicity.

Definition 4.7.9 A Hilbert smooth game is *monotone* if, for all $(s, t) \in S \times S$,

$$\langle \vec{\nabla} g(s) - \vec{\nabla} g(t), s - t \rangle \leqslant 0,$$

and the game is strictly monotone if the inequality is strict whenever $s \neq t$.

This monotonicity condition was introduced by Rosen [178].

Theorem 4.7.10 *For a Hilbert smooth monotone game:*

(1) A profile $s \in S$ is a Nash equilibrium if and only if $\forall t \in S$, $\langle \vec{\nabla} g(t), s - t \rangle \geqslant 0$.
(2) If the game is strictly monotone, a Nash equilibrium is unique.

Proof (1) By monotonicity and the characterization in Theorem 4.7.6, if s is a Nash equilibrium then for all $t \in S$:

$$0 \leqslant \langle \vec{\nabla} g(s), s - t \rangle \leqslant \langle \vec{\nabla} g(t), s - t \rangle.$$

Conversely, suppose $\langle \vec{\nabla} g(t), s - t \rangle \geqslant 0$ for all $t \in S$, or equivalently, for all $i \in I$ and z^i in S^i, $\langle \nabla g(z^i, s^{-i}), s^i - z^i \rangle \geqslant 0$ (by taking $t = (z^i, s^{-i})$). Now fix a player i and a deviation t^i. By the mean value theorem, there is a $z^i = \lambda t^i + (1 - \lambda)s^i$ such that $g^i(s^i, s^{-i}) - g^i(t^i, s^{-i}) = \langle \nabla g(z^i, s^{-i}), s^i - t^i \rangle$ and $s^i - z^i = \lambda(s^i - t^i)$. Consequently, $g^i(s^i, s^{-i}) \geqslant g^i(t^i, s^{-i})$: s is a Nash equilibrium.

(2) Let s and t be two Nash equilibria. By the characterization in Theorem 4.7.6, $\langle \vec{\nabla} g(s), s - t \rangle \geqslant 0$ and $\langle \vec{\nabla} g(t), t - s \rangle \geqslant 0$. This implies that $\langle \vec{\nabla} g(s) - \vec{\nabla} g(t), s - t \rangle \geqslant 0$. By strict monotonicity, we have the opposite inequality, hence equality, thus $s = t$.   □

4.8 Discontinuous Games

4.8.1 Reny's Solution in Discontinuous Games

In this and the next two subsections, payoff functions may be discontinuous but strategy sets are assumed to be convex subsets of a topological vector space and the game is assumed to be compact and quasi-concave. For example, consider a Nash bargaining game where players seek to divide an amount M among them. Simultaneously, each player $i \in I$ proposes a number $x^i \in [0, 1]$. If the proposal is feasible (i.e. $\sum_{i \in I} x^i \leqslant 1$), each player $i \in I$ gets the payoff $x^i M$. Otherwise, all players get a payoff of 0. Here there is a discontinuity of i's payoff function $y^i \mapsto g^i(y^i, x^{-i})$ at $x^i = 1 - \sum_{j \neq i} x^j$.

In a timing game, each player $i \in I$ chooses a time $t^i \in [0, 1]$. The game stops at the first time $\theta = \min_{i \in I} t^i$ when a player acts and the payoff function g^i of player i depends on his identity, the stopping time θ, and the stopping coalition $S = \arg \min_{i \in I} t^i$. Here there is a discontinuity of $s^i \mapsto g^i(s^i, t^{-i})$ at $t^i = \theta$.

Reny [174] provided a general discontinuity condition (called better reply security) that proves to be satisfied in many discontinuous games. This and the next two subsections explore Reny's approach (presented slightly differently, see [24]).

Definition 4.8.1 The *graph* of G is the set $\Gamma = \{(s, v) \in S \times \mathbb{R}^N : v = g(s)\}$. Denote by $\overline{\Gamma}$ its closure.

Since the game is compact, we deduce that $\overline{\Gamma}$ is a compact subset of $S \times \mathbb{R}^N$. For all i and $s \in S$, we introduce the lower semi-continuous regularization of g^i with respect to its variable s^{-i} as follows:

$$\underline{g}^i(s^i, s^{-i}) = \sup_{U \ni s^{-i}} \inf_{\widetilde{s}^{-i} \in U} g^i(\widetilde{s}^i, s^{-i}) = \lim_{s_n^{-i} \to s^{-i}} \inf g^i(s^i, s_n^{-i}),$$

where the supremum is taken over all open neighbourhoods of s^{-i}. By construction, $\underline{g}^i(s^i, s^{-i}) \leqslant g^i(s^i, s^{-i})$, and $s^{-i} \to \underline{g}^i(s^i, s^{-i})$ is lower semi-continuous.

Definition 4.8.2 A pair $(\overline{s}, \overline{v}) \in \overline{\Gamma}$ (strategy profile, payoff profile) is a *Reny solution* if for every $i \in I$, $\sup_{s^i \in S^i} \underline{g}^i(s^i, \overline{s}^{-i}) \leqslant \overline{v}^i$.

Theorem 4.8.3 ([174]) *Any compact quasi-concave game G admits a Reny solution and the set of Reny solutions is compact.*

It is instructive to compare this proof with that of Theorem 4.7.2.

Proof Given any $(s, t) \in S \times S$, define

$$\underline{g}(s, t) = \left(\underline{g}^1(s^1, t^{-1}), \ldots, \underline{g}^i(s^i, t^{-i}), \ldots, \underline{g}^N(s^N, t^{-N}) \right)$$

and $E(s) = \left\{ (t, h) \in \overline{\Gamma} : \underline{g}(s, t) \leqslant h \right\}$. Thus, a Reny solution exists if and only if $\bigcap_{s \in S} E(s)$ is non-empty. Since $\underline{g}^i(s^i, t^{-i})$ is lower semi-continuous in t^{-i} for every i, $E(s)$ is compact for all $s \in S$. Consequently, to prove that $\bigcap_{s \in S} E(s)$ is non-empty, it is sufficient to show that $\bigcap_{s \in S_0} E(s)$ is non-empty for every finite subset $S_0 = \{s_1, \ldots, s_m\}$ of S, or equivalently that $\exists (t, h) \in \overline{\Gamma}$ such that $\underline{g}(s, t) \leqslant h$ for all $s \in S_0$.

Define $S_0^i = \{s_1^i, \ldots, s_m^i\}$ and let $\mathrm{co}S_0^i$ denote the convex hull of S_0^i. Since each S_0^i is finite, we may endow its convex hull with the Euclidean metric. Because each S^i is a subset of a topological vector space, the topology S_0^i induced by the Euclidean metric is at least finer than the original one (they coincide when S^i is Hausdorff). Hence, convergence with respect to the Euclidean metric implies convergence with respect to the original topology.

Restricted to the vector space spanned by $\mathrm{co}S_0^i$, $i \in I$, we obtain strategy sets that are subsets of a finite-dimensional topological vector space where there is only one possible topology, and such a space may be equipped with a norm compatible with this topology (all norms are equivalent). Since $\underline{g}^i(s^i, \cdot)$ is lower-semi-continuous on $\prod_{j \neq i} \mathrm{co}S_0^j$, we can apply the following lemma to approximate it from below by a sequence of continuous functions.

Lemma 4.8.4 ([37, 174]) *Let Y be a compact metric space and $f : Y \to \mathbb{R}$ be lower-semi-continuous. Then there exists a sequence of continuous functions $f_n : Y \to \mathbb{R}$ such that, for all $y \in Y$:*

(i) $f_n(y) \leqslant f(y)$.
(ii) $\forall y_n \to y : \liminf_n f_n(y_n) \geqslant f(y)$.

Proof Let $F = \{g \in \mathcal{C}(Y) : g(y) \leqslant f(y), \ \forall y \in Y\}$, where $\mathcal{C}(Y)$ is the metric space of bounded continuous functions endowed with the maximum metric. Because F is closed and $\mathcal{C}(Y)$ is separable, F is separable and so there is a countable dense subset $\{g_1, g_2, \ldots\}$ of F. Define $f_n(y) = \max\{g_1(y), \ldots, g_n(y)\}$ for all $y \in Y$. Then the sequence f_n is in $\mathcal{C}(Y)$ and satisfies (i). It remains to show that it satisfies (ii). Suppose the contrary, and let $y_n \mapsto y^0$, $f_n(y_n) \mapsto \alpha < f(y^0)$. Thus, $f(y^0) > \alpha + \varepsilon$ for some $\varepsilon > 0$. Because f is lower semi-continuous, one can prove that there is an $h \in F$ and a neighborhood U of y^0 such that $h(y) > \alpha + \varepsilon$ for all $y \in U$. Since $\{g_1, g_2, \ldots\}$ is dense in F, there is a k such that g_k is within $\varepsilon/2$ of h. Thus, for all $n \geqslant k$ such that $y_n \in U$ one has $f_n(y_n) \geqslant g_k(y_n) \geqslant h(y_n) - \varepsilon/2 \geqslant \alpha + \varepsilon/2$, a contradiction. $\square$

Consequently, for all $s^i \in S_0^i$, there is a sequence of continuous functions $g_n^i(s^i, \cdot)$ defined on $\prod_{j \neq i} \mathrm{co}S_0^j$ such that for all $s^{-i} \in \prod_{j \neq i} \mathrm{co}S_0^j$:

(i) $g_n^i(s^i, s^{-i}) \leqslant \underline{g}^i(s^i, s^{-i})$.
(ii) $\forall s_n^{-i} \to s^{-i} : \liminf_n g_n^i(s^i, s_n^{-i}) \geqslant \underline{g}^i(s^i, s^{-i})$.

Now construct a sequence of games G_n as follows. The set of pure strategies of each player i in G_n is $\Delta(S_0^i)$. For all $\mu \in \prod_{j \in I} \Delta\left(S_0^j\right)$, the payoff function of player i in G_n is

$$f_n^i(\mu) = \sum_{s^i \in S_0^i} g_n^i(s^i, \bar{s}^{-i}) \mu^i(s^i),$$

where $\bar{s}^j = \bar{s}^j(\mu^j) = \sum_{s^j \in S_0^j} \mu^j(s^j) s^j \in \text{co} S_0^j$. Each game G_n satisfies the assumptions of Theorem 4.7.2, it admits a Nash equilibrium μ_n (and an associated $\bar{s}_n$, where $\bar{s}_n^j = \sum_{s^j \in S_0^j} \mu_n^j(s^j) s^j \in \text{co} S_0^j$). By linearity of f_n^i in μ^i, for all i and all $s^i \in S_0^i$ such that $\mu_n^i(s^i) > 0$ and all $\tilde{s}^i \in S_0^i$,

$$g_n^i(\tilde{s}^i, \bar{s}^{-i}) \leqslant f_n^i(\mu_n) = g_n^i(s^i, \bar{s}_n^{-i}) \leqslant \underline{g}^i(s^i, \bar{s}_n^{-i}) \leqslant g^i(s^i, \bar{s}_n^{-i}).$$

The first inequality and equality are consequences of the facts that μ_n is a Nash equilibrium of G_n and that the payoff function $f_n^i(\mu^i, \mu^{-i})$ is linear in μ^i (hence all $s^i \in \text{supp}(\mu_n^i)$ are best responses to μ_n^{-i} in G_n). The second inequality above is a consequence of (i) in Lemma 4.8.4. Integrating with respect to μ_n^i and using the quasi-concavity of g^i in s^i, we obtain that, for all $i \in I$ and $\tilde{s}^i \in S_0^i$,

$$g_n^i(\tilde{s}^i, \bar{s}_n^{-i}) \leqslant f_n^i(\mu_n) \leqslant g^i(\bar{s}_n).$$

By compactness, without loss of generality one can assume that $\bar{s}_n \to \bar{s}$ and that $g^i(\bar{s}_n) \to \bar{v}^i$. Thus, for all $\tilde{s} \in S_0$,

$$\underline{g}_i(\tilde{s}_i, \bar{s}_{-i}) \leqslant \liminf_n g_n^i(\tilde{s}^i, \bar{s}_n^{-i}) \leqslant \bar{v}^i,$$

the first inequality follows from (ii) in Lemma 4.8.4. Hence $(\bar{s}, \bar{v}) \in E(s) \forall s \in S_0$. $\qquad \square$

4.8.2 Nash Equilibria in Discontinuous Games

Recall that in this section, strategy sets are assumed to be convex subsets of a topological vector space and the game is assumed to be compact and quasi-concave.

Definition 4.8.5 A game G is *better reply secure* if for all Reny solutions $(\bar{s}, \bar{v}) \in \Gamma$, $\bar{s}$ is a Nash equilibrium.

Since whenever $\bar{s}$ is a Nash equilibrium, $(\bar{s}, g(\bar{s}))$ is obviously a Reny solution, a game is better reply secure if Nash profiles and Reny solution profiles coincide.

Corollary 4.8.6 ([174]) *If a game is compact quasi-concave and better reply secure, its set of Nash equilibria is non-empty and compact.*

Existence is obvious because the set of Reny solutions is non-empty and compact. Since any continuous game is better reply secure, Corollary 4.8.6 extends Theorem 4.7.2.

Let us provide now conditions on the data of the game that imply better reply security.

Definition 4.8.7 (i) G is *payoff secure* if for all $i \in I$ and all $s^{-i} \in S^{-i}$:

$$\sup_{t^i \in S^i} g^i(t^i, s^{-i}) = \sup_{t^i \in S^i} \underline{g}^i(t^i, s^{-i}).$$

(ii) G is *reciprocal upper semi-continuous* if for all $(\overline{s}, \overline{v}) \in \overline{\Gamma}$:

$$\text{if } g^i(\overline{s}) \leqslant \overline{v}^i \text{ for every } i \in I \text{ then } g(\overline{s}) = \overline{v}.$$

Alternatively, let us say that player i can *secure a payoff* α^i at $s \in S$ if there exists an $\overline{s}^i \in S^i$ s.t. $g^i(\overline{s}^i, \tilde{s}^{-i}) \geqslant \alpha^i$ for all $\tilde{s}^{-i}$ in an open neighbourhood U containing s^{-i}. Thus, G is payoff secure if for all $s \in S$ and all $\varepsilon > 0$, each player $i \in I$ can secure a payoff above $g^i(s) - \varepsilon$.

Observe that if $\sum_{i \in I} g^i$ is continuous, the game is reciprocal upper semi-continuous. Thus, when g^i is lower semi-continuous in s^{-i} for every i and the sum of payoffs is constant, the game is payoff secure and reciprocal upper-semi-continuous.

Corollary 4.8.8 *Every payoff secure and reciprocal upper semi-continuous game is better reply secure.*

This extends Sion's theorem when strategy sets are compact.

Proof Let $(\overline{s}, \overline{v}) \in \overline{\Gamma}$ be a Reny solution. Thus, for every $i \in I$, $\sup_{t^i \in S^i} \underline{g}^i(t^i, \overline{s}^{-i}) \leqslant \overline{v}^i$. Since the game is payoff secure, $\sup_{t^i \in S^i} g^i(t^i, \overline{s}^{-i}) = \sup_{t^i \in S^i} \underline{g}^i(t^i, \overline{s}^{-i})$. The two inequalities together imply that $g^i(\overline{s}) \leqslant \overline{v}^i$ for every $i \in I$. By reciprocal upper semicontinuity, $g(\overline{s}) = \overline{v}$. Consequently, $\overline{s}$ is a Nash equilibrium. $\square$

4.8.3 Approximate Equilibria in Discontinuous Games

Recall that in this subsection, strategy sets are assumed to be convex subsets of a topological vector space and the game is assumed to be compact and quasi-concave.

Definition 4.8.9 A pair $(\overline{s}, \overline{v}) \in \overline{\Gamma}$ (strategy profile, payoff profile) is an *approximate solution* (and $\overline{s}$ is an *approximate equilibrium*) if there exists a sequence $(s_n)_n$ in S and a sequence $(\varepsilon_n)_n$ of positive real numbers, converging to 0, such that:

(i) for every n, s_n is an ε_n-equilibrium;
(ii) the sequence $(s_n, g(s_n))$ converges to $(\overline{s}, \overline{v})$.

In zero-sum games, the existence of an approximate solution implies the existence of a value.

Proposition 4.8.10 *Any approximate solution is a Reny solution.*

Proof Let $(s_n)_{n\in\mathbb{N}}$ be a sequence of ε_n-equilibria such that $(s_n, g(s_n))$ converges to $(\overline{s}, \overline{v})$. By definition, $g^i(t^i, s_n^{-i}) \leqslant g^i(s_n) + \varepsilon_n$ for every n, every player $i \in I$ and every deviation $t^i \in S^i$. Taking the infimum limit as n tends to infinity leads to $\underline{g}^i(t^i, \overline{s}^{-i}) \leqslant \overline{v}^i$. Thus, $(\overline{s}, \overline{v})$ is a Reny solution. □

Definition 4.8.11 A game G is *approximately better reply secure* if for all Reny solutions $(\overline{s}, \overline{v}) \in \overline{\Gamma}$, $\overline{s}$ is an approximate equilibrium.

Corollary 4.8.12 ([24]) *Every approximately better reply secure quasi-concave and compact game admits an approximate equilibrium.*

Proof The set of Reny solutions is non-empty. □

Definition 4.8.13 A game G has the *marginal continuity property* if for every $s \in S$ and $i \in I$, $s^{-i} \mapsto \sup_{t^i \in S^i} g^i(t^i, s^{-i})$ is continuous.

Corollary 4.8.14 ([163]) *Every payoff secure compact game with the marginal continuity property is approximately better reply secure.*

Proof If $(\overline{s}, \overline{v})$ is a Reny solution, then

$$\sup_{t^i \in S^i} \underline{g}^i(t^i, \overline{s}^{-i}) = \sup_{t^i \in S_i} g^i(t^i, \overline{s}^{-i}) \leqslant \overline{v}^i,$$

the equality being a consequence of payoff security. Since $\overline{v} = \lim_{s^n \to s} g(s_n)$ for some sequence s_n, the continuity of $\sup_{t^i \in S_i} g^i(t^i, \overline{s}^{-i})$ at $\overline{s}^{-i}$ guarantees that $(\overline{s}, \overline{v})$ is an approximate solution. □

Example 4.8.15 **First-price auction** Two players $i = 1, 2$ compete for a good. They make a bid $x^i \in [0, 1]$, and receive a payoff:

$$g^i(x^i, x^j) = \begin{cases} w^i - x^i & \text{if } x^i > x^j, \\ \frac{w^i - x^i}{2} & \text{if } x^i = x^j, \\ 0 & \text{if } x^i < x^j, \end{cases}$$

where $w^1 \in (0, 1)$ is the valuation of the good for player 1 and $w^2 \in [0, w^1)$ is the valuation for player 2. This game has no Nash equilibrium. However, it is payoff secure, compact, and has the marginal continuity property. Every $(x_1, x_2, v_1, v_2) = (y, y, w_1 - y, 0)$, where $y \in [w_2, w_1]$ is an approximate (= Reny) solution. They are limit points of ε-equilibria, $\varepsilon > 0$, in which player 1 bids $y + \varepsilon$ and player 2 bids $y \in [w_2, w_1]$.

4.9 Semi-algebricity of the Set of Nash Equilibria

This section is devoted to the study of finite games. Each player i's pure strategy set S^i has cardinality m^i. Let $m = \Pi_i m^i$. Then every game may be identified with a point in $\mathbb{R}^{Nm}$. For example, any two-player game where each player has two strategies can be mapped to $\mathbb{R}^8$:

	L	R
T	(a_1, a_2)	(a_3, a_4)
B	(a_5, a_6)	(a_7, a_8)

Proposition 4.9.1 *The set of mixed Nash equilibria of a finite game is defined by finitely many polynomial weak inequalities.*

Proof We use the linearity of payoff functions with respect to each player's mixed strategy to reduce the equilibrium test against finitely many pure deviations. Thus $\sigma = \{\sigma^i(s^i)\} \in \mathbb{R}^{\sum_{i \in I} m^i}$ is a mixed Nash equilibrium if and only if it is the solution of the following set of polynomial weak inequalities:

$$\sum_{s^i \in S^i} \sigma^i(s^i) - 1 = 0, \quad \sigma^i(s^i) \geqslant 0, \quad \forall s^i \in S^i, \forall i \in I,$$

and

$$\sum_{s=(s^1,\dots,s^N) \in S} \prod_{j \in I} \sigma^j(s^j) g^i(s)$$

$$\geqslant \sum_{s^{-i} \in S^{-i}} \prod_{j \in I, j \neq i} \sigma^j(s^j) g^i(t^i, s^{-i}), \forall t^i \in S^i, \forall i \in I.$$

$\square$

A closed set in $\mathbb{R}^k$ is *semi-algebraic* if it is the finite union of sets of the form $\{x : P_k(x) \geqslant 0, \forall k = 1, \dots, r\}$, where each P_k is a polynomial.

Theorem 4.9.2 *Any semi-algebraic closed set of $\mathbb{R}^k$ has finitely many connected components.*

See Benedetti and Risler [18].

Corollary 4.9.3 *The set of Nash equilibria of a finite game has finitely many connected components, which are semi-algebraic closed sets. These are called the Nash components of the game.*

To illustrate this semi-algebraic structure, consider the following example:

	L	M	R
T	(2, 1)	(1, 0)	(1, 1)
B	(2, 0)	(1, 1)	(0, 0)

It has a unique Nash component, which is homeomorphic to a segment.

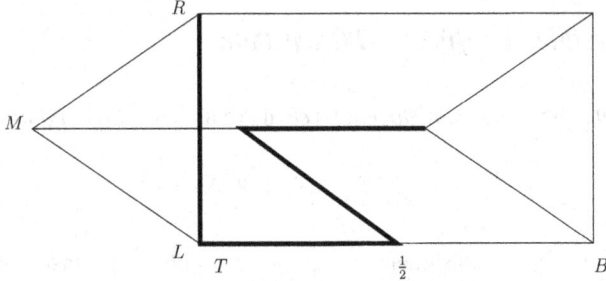

In the following example (from Kohlberg and Mertens [108]) there is a unique Nash component, homeomorphic to a circle.

	L	M	R
T	(1, 1)	(0, −1)	(−1, 1)
m	(−1, 0)	(0, 0)	(−1, 0)
B	(1, −1)	(0, −1)	(−2, −2)

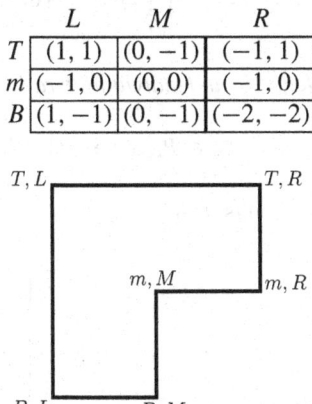

Interestingly, any equilibrium point in this Nash component may be obtained as the limit of the unique equilibrium of a nearby game. For example, let $\varepsilon > 0$ be small and consider the following payoff-perturbation of the game:

	L	M	R
T	$(1, 1 − \varepsilon)$	$(\varepsilon, −1)$	$(−1 − \varepsilon, 1)$
m	$(−1, −\varepsilon)$	$(−\varepsilon, \varepsilon)$	$(−1 + \varepsilon, −\varepsilon)$
B	$(1 − \varepsilon, −1)$	$(0, −1)$	$(−2, −2)$

This game has a unique mixed Nash equilibrium where player 1 plays $(\varepsilon/(1 + \varepsilon), 1/(1 + \varepsilon), 0)$ and player 2 plays $(0, 1/2, 1/2)$. As ε goes to zero, it converges to the equilibrium of the original game: $(0, 1, 0)$ for player 1 and $(0, 1/2, 1/2)$ for player 2.

4.10   Complements

4.10.1   Feasible Payoffs and Threat Point

Let G be a finite game. The *punishment level*, also called the *threat point*, for player i is the quantity

$$V^i = \min_{\sigma^{-i} \in \prod_{j \neq i} \Delta(S^j)} \max_{s^i \in S^i} g^i(s^i, \sigma^{-i}).$$

This is the maximal punishment players $-i$ can enforce against player i using mixed strategies (without correlation). V^i plays an important rôle in repeated games (the Folk theorem), as will be seen in the last Chap. 8. The punishment vector payoff is $V = (V^i)_{i \in I}$. The set of *feasible payoffs* in the one-shot game is

$$P_1 = \{x \in \mathbb{R}^n; \exists \sigma \in \prod_{i \in I} \Delta(S^i), G(\sigma) = x\}.$$

Consequently, R_1, the *set of feasible and individually rational payoffs* in the one-shot game, is

$$R_1 = \{x \in P_1; x^i \geqslant V^i\}.$$

Thus, any Nash equilibrium belongs to R_1.
Here are three examples:

	L	R
T	(1, 1)	(1, 0)
B	(0, 1)	(0, 0)

	L	R
T	(1, 1)	(0, 1)
B	(1, 0)	(0, 0)

	L	R
T	(3, 1)	(0, 0)
B	(0, 0)	(1, 3)

One can show that R_1 is closed and arc-connected. It is contractible in two-player games, but not for three players or more (see Exercise 3). When the game is infinitely repeated, the set of feasible payoffs increase to $R_\infty = coR_1 = \{x \in \mathbb{R}^n; \exists \sigma \in \Delta(S), G(\sigma) = x\}$ (see Chap. 8).

4.10.2 Invariance, Symmetry, Focal Points and Equilibrium Selection

How do we select between many Nash equilibria? A selection procedure has to satisfy some natural properties. One desirable axiom [108] is BR-invariance. It states that two games with the same best-response correspondence must have the same set of solutions. The set of Nash equilibria and the set of rationalizable strategies are BR-invariant. What could the other possible axioms be?

A *focal point* (also called a Schelling point) is a Nash equilibrium that people will play in the absence of communication, because it is natural, special, or relevant to them. A symmetric or a Pareto dominating equilibrium could be considered as a natural focal point (an equilibrium payoff v Pareto dominates w if $v^i \geqslant w^i$ for all $i \in I$ and there is at least one strict inequality).

To understand the incompatibility between the different principles, consider the following symmetric game:

	L	R
T	(3, 3)	(0, 0)
B	(0, 0)	(2, 2)

Let a, b, c, d be any real numbers. Transform the above matrix as follows:

	L	R
T	(3 + a, 3 + c)	(0 + b, 0 + c)
B	(0 + a, 0 + d)	(2 + b, 2 + d)

Since the best response correspondence is the same in both games, BR-invariance implies that the two games should have the same solution. If $a = c = -1$ and $b = d = 1$, we obtain the game:

	L	R
T	(2, 2)	(1, −1)
B	(−1, 1)	(3, 3)

and if $a = c = -1$ and $b = d = 0$, the game is:

	L	R
T	(2, 2)	(0, −1)
B	(−1, 0)	(2, 2)

All these games have two pure Nash equilibria (T, L) and (B, R) and one mixed: $[(\frac{2}{5}T + \frac{3}{5}B), (\frac{2}{5}L + \frac{3}{5}R)]$. If a selection theory predicts (T, L) in the first game (a natural focal point because players have a clear common interest to do so), then this theory will violate BR-invariance (it predicts (B, R) when $a = c = -1$ and $b = d = 1$). On the other hand, if a selection theory should respect all symmetries of a game, then it predicts the mixed equilibrium in the last example ($a = c = -1$ and $b = d = 0$).

4.10.3 Nash Versus Prudent Behavior

A strategy $\sigma^i \in \Delta(S^i)$ of player i in a finite game G is prudent if for all $s^{-i} \in S^{-i}$:

$$g^i(\sigma^i, s^{-i}) \geq \max_{\tau^i \in \Delta(S^i)} \min_{t^{-i} \in S^{-i}} g^i(\tau^i, t^{-i}) := \underline{V}^i.$$

Playing a prudent strategy guarantees to player i at least $\underline{V}^i$. A prudent behavior makes sense if a player knows nothing about the other players' payoff functions, the other players' information or rationality. In computer science, this notion is very popular. But what happens if the game and rationality are common knowledge among the players? Should they necessarily play according to the unique Nash equilibrium recommendation? Consider the following example [12]:

	L	R
T	(2, 0)	(0, 1)
B	(0, 1)	(1, 0)

The unique Nash equilibrium is $\tau = (\frac{1}{2}T + \frac{1}{2}B; \frac{1}{3}L + \frac{2}{3}R)$ with a payoff $\frac{2}{3}$ to player 1 and $\frac{1}{2}$ to player 2. The crucial property of this game is that the Nash payoff coincides with the prudent payoff $\underline{V}$.

Thus, if player 1 is rational (in the sense of Savage: he maximizes his payoff given his belief) and if he expects player 2 to play $\tau^2 = \frac{1}{3}L + \frac{2}{3}R$ then any strategy $xT + (1 - x)B$ is rational and they all yield him the same expected payoff $\frac{2}{3}$. Why choose τ^1 and not another strategy? Player 1 can equally optimally decide to play his prudent strategy $\sigma^1 = \frac{1}{3}T + \frac{2}{3}B$, which guarantees the payoff $\frac{2}{3}$. However, if the *social convention* is to play prudently, and if the players know it, then player 1 would deviate to T.

4.10.4 The Impact of Common Knowledge of the Game

For a profile of strategies to be "stable" against individual deviations, hence a Nash equilibrium, players do not need to know the payoff functions of the opponents, only their behavior (what they intend to play). However, when payoff functions are common knowledge, some Nash equilibria could be ruled out. Consider the following example:

	L	R
T	(1, 1)	(0, 1)
B	(0, 0)	(2, 0)

Here, player 2's payoff functions do not depend on his own actions, but do depend on player 1's actions. Thus, any choice of player 2 is rational: no deviation is profitable. The set of Nash equilibria is given by the best response correspondence of player 1.

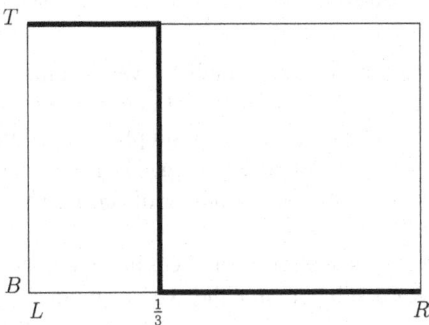

Imagine now that payoff functions and rationality are common knowledge among the players. In that case, when player 2 plays L, the unique best response of player 1 is to play T, yielding to player 2 his best possible payoff in the game. Hence, player 2, knowing player 1's best response correspondence, can anticipate the rational behavior of player 1 and can credibly force him to choose T by choosing L because if player 1 deviates to B, he strictly loses, while the action L is always rational for player 2 whatever player 1 plays, and player 1 knows it (this reasoning uses, among other things, that player 1 knows that player 2 knows player 1's payoff function). Thus, if the game and rationality are common knowledge, we should expect (T, L) to be played. In fact, one can observe that the above reasoning needs only two levels of knowledge.

However, sometimes common knowledge of rationality and of the game cannot help removing equilibria even if players can communicate before the play.

	L	R
T	(7, 7)	(0, 6)
B	(6, 0)	(5, 5)

In the above example [5], even if players try to convince each other that they have a common interest to play (T, L), this does not help. Whatever player 1 intends to play, he prefers that player 2 plays L and symmetrically, whatever player 2 intends to play, he prefers player 1 to play T. Thus, if the norm in this society is to play (B, R), no player can credibly convince the other to change, because every player has an interest that the other player changes even if he does not change, and players know it (because the game is common knowledge).

To predict (T, L) as the outcome in this example, other principles should be assumed in addition to common knowledge of rationality and of the game, such as: if in a game some Nash profile Pareto dominates all other profiles, it must be played.

We will see in Chap. 6 that in extensive form games, common knowledge of the game and of rationality help to refine the set of Nash equilibria, via the concepts of backward and forward inductions.

4.11 Fixed Point Theorems

To prove Brouwer's theorem, we first establish a very useful lemma due to Sperner. Let Δ be a unit simplex of dimension k, and denote its $k + 1$ extreme points by $\{x^0, \ldots, x^k\}$, hence $\Delta = \text{co}\{x^0, \ldots, x^k\}$. A simplicial subdivision of Δ is a finite collection of sub-simplices $\{\Delta_i\}$ of Δ satisfying (1) $\bigcup_i \Delta_i = \Delta$ and (2) for all (i, j), $\Delta_i \cap \Delta_j$ is empty or is some sub-simplex of the collection. The mesh of a subdivision is the largest diameter of a sub-simplex.

Let V be the set of vertices of all sub-simplices in $\{\Delta_i\}$. Each vertex $v \in V$ decomposes as a unique convex combination of the extreme points: $v = \sum_{i=0}^{k} \alpha^i(v) x^i$. Let

$$I(v) = \{i : \alpha^i(v) > 0\} \subset \{0, \ldots, k\}$$

be the set of indices with a strictly positive weight in the decomposition (support of v). Thus, $\text{co}\{x^i : i \in I(v)\}$ is the minimal face of Δ that contains v in its relative interior.

A *labeling* of V is a function that associates to each $v \in V$ an integer in $I(v)$. There are $k + 1$ possible labels, an extreme point x^j has the label j, and a point v in the interior of the face $\text{co}\{x^{i_1}, \ldots, x^{i_m}\}$ has one of the labels $\{i_1, \ldots, i_m\}$.

A sub-simplex Δ_i is called *completely labeled* if its vertices (extreme points) have all the $k + 1$ labels.

Lemma 4.11.1 ([206]) *Every labeling of any simplicial subdivision of a simplex Δ has an odd number of completely labeled sub-simplices.*

Proof The proof is by induction on the dimension k of the simplex Δ. For $k = 0$, this is trivial. Suppose the result holds for dimension $k - 1$ and let us prove that any labeling in dimension k has an odd number of sub-simplices.

Imagine that Δ is a house, and that any sub-simplex Δ_i with full dimension is a room. A door is a sub-simplex of dimension $k - 1$ having exactly the labels 0 to $k - 1$. Thus, a room Δ_i has no doors, 1 door or 2 doors. Actually, suppose that k vertices of Δ_i have the labels $0, \ldots, k - 1$. If the last vertex has the missing label k, then the room has all labels (and only one door). Otherwise, the room has two doors (and label k is missing). The induction hypothesis implies that there is an odd number of doors in the face $F = \text{co}\{x^0, \ldots, x^{k-1}\}$ of $\Delta = \text{co}\{x^0, \ldots, x^k\}$. Let us show that there is an odd number of rooms having all the labels.

The proof is algorithmic. Imagine that you enter the house Δ from outside using some door in the face F. If the room has another door, take it and keep going, until (1) you reach a room without other doors (and this would be a completely labeled room) or (2) you leave the house by the face F (because, by definition of a labeling, it is the unique face having doors). Note that no cycle can occur since there is no first room visited twice. Thus, linked doors on the face F go by pairs. Since there is an odd number of doors in F (by the induction hypothesis), there is an odd number of completely labeled rooms that may be reached from outside.

Finally, let Δ_i be a completely labeled room that cannot be reached from outside. This room has a door, which leads to another room, and so on until (1) one reaches a completely labeled room or (2) one leaves the house by the face F. But (2) is impossible because it implies that Δ_i can be reached from outside. Consequently, completely labeled rooms that cannot be reached from outside go by pairs. Thus, Δ has an odd number of completely labeled rooms. $\qquad \square$

There is a stronger version of Sperner's lemma: the number of completely labeled sub-simplices with a positive orientation (the same as Δ) is exactly one more than the number of completely labeled sub-simplices with a negative orientation. The proof is quite similar: one shows that two completely labeled linked rooms have opposite orientation, and proceed by induction for rooms that are connected to the outside.

Corollary 4.11.2 *Every continuous function $f : \Delta \to \Delta$ from a simplex to itself has a fixed point.*

Proof Let $\Delta = \{(r^0, \ldots, r^k) \in \mathbb{R}^{k+1}; r^i \geqslant 0, \sum_{i=0}^{k} r^i = 1\}$ be the unit simplex of dimension k and fix $\varepsilon > 0$. Consider any simplicial subdivision of Δ with a mesh smaller than ε and let λ be a labeling of V satisfying

$$\lambda(v) \in I(v) \cap \{i : f^i(v) \leqslant v^i\}.$$

λ is well defined. Indeed, if the above intersection is empty, then $1 = \sum_{i=0}^{k} v^i = \sum_{i \in I(v)} v^i < \sum_{i \in I(v)} f(v)^i \leqslant \sum_{i=0}^{k} f(v)^i = 1$, a contradiction.

By Sperner's lemma, there is a completely labeled sub-simplex Δ_ε (with diameter smaller than ε), whose $k+1$ extreme points $\{v(i, \varepsilon), i = 0, \ldots, k\}$ satisfy $f(v(i, \varepsilon))^i \leqslant v(i, \varepsilon)^i$ for all $i = 0, \ldots, k$, and $\|v(i, \varepsilon) - v(j, \varepsilon)\| \leqslant \varepsilon$, for all i and j. Using compactness and letting ε go to zero leads to the existence of a $v \in \Delta$ satisfying, for all $i = 0, \ldots, k$, $f(v)^i \leqslant v^i$. But since $\sum_{i=0}^{k} v^i = \sum_{i=0}^{k} f(v)^i = 1$, we obtain $f(v) = v$. $\qquad \square$

Lemma 4.11.3 *Assume that K is a retract of the simplex Δ. Then, any continuous function $f : K \to K$ has a fixed point.*

Proof Let h be a retraction from Δ to K. h is a continuous map from Δ to $K \subset \Delta$ and $h|_K = \text{Id}|_K$. Then $f \circ h : \Delta \to \Delta$ is continuous and has a fixed point z. But $z \in K$, thus $f \circ h(z) = f(z) = z$ is a fixed point of f in K. $\qquad \square$

Theorem 4.11.4 ([32]) *Let C be a non-empty convex and compact subset of a finite-dimensional Euclidean space. Then any continuous function $f : C \to C$ has a fixed point.*

Proof Let Δ be a simplex that contains C. Since C is convex and closed, Π_C, the projection on C, is a retraction from Δ to C. The previous lemma applies. □

This theorem can be extended to correspondences as follows.

Theorem 4.11.5 ([104]) *Let C be a non-empty convex and compact subset of a normed vector space and let F be a correspondence from C to C such that:*

(i) For all $c \in C$, $F(c)$ is convex compact and non-empty;
(ii) The graph $\Gamma = \{(c,d) \in C \times C : d \in F(c)\}$ of F is closed.

Then, $\{c \in C : c \in F(c)\}$ is non-empty and compact.

Proof Suppose F has no fixed point and let $\delta(x)$ be the distance from x to the non-empty compact convex set $F(x)$. Then $\delta(x) > 0$ for all $x \in C$. Let $\Omega_x = \{y \in C; d(y, F(x)) < \delta(x)/2\}$, which is open and convex. Also, $U_x = \{z \in C; F(z) \subset \Omega_x\}$ is an open set that contains x (because the graph of F is closed), consequently it contains an open ball $B(x, s(x))$ with $0 < s(x) < \delta(x)/3$. Thus $B(x, s(x)) \cap \Omega_x = \varnothing$. The open balls $B(x, s(x)/2)$ form an open cover of the compact C, thus there exists a finite sub-cover $\{B(x_i, s(x_i)/2)\}_i$. Let $r = \min_i s(x_i)/2$. Then for all $z \in C$ the open ball $B(z, r)$ is included in $B(x_i, s(x_i))$ for some i. One can thus extract from the covering of C by $\{B(z, r)\}_z$ a new finite subcovering $\{B(z_k, r)\}_{k \in K}$. Define for all k and $x \in C$ the continuous function

$$f_k(x) = \frac{d(x, B(z_k, r)^c)}{\sum_j d(x, B(z_j, r)^c)},$$

where $B(z_k, r)^c$ denotes the complement of $B(z_k, r)$ in C. This forms a partition of unity: f_k is continuous from C to $[0, 1]$, $\sum_k f_k = 1$ and for all k, $f_k(x) = 0$ for $x \notin B(z_k, r)$. Let $y_k \in F(z_k)$ for every k and define $f(x) = \sum_k f_k(x) y_k$. Then f is a continuous function from co $\{y_k, k \in K\}$ to itself. By Brouwer's theorem, f has a fixed point x. If $f_k(x) > 0$ then $x \in B(z_k, r)$ and so $z_k \in B(x, r) \subset B(x_i, s(x_i)) \subset U_{x_i}$ for some i (which depends on x but not on k). Thus, $y_k \in F(z_k) \subset \Omega_{x_i}$ for every k and by convexity of Ω_{x_i} we deduce that $f(x) \in \Omega_{x_i}$. But $x \in B(x_i, s(x_i))$ and $\Omega_{x_i} \cap B(x_i, s(x_i)) = \varnothing$ contradicts $f(x) = x$. □

This theorem was extended by Glicksberg [79] and Fan [53] to any Hausdorff locally convex topological vector space. It is remarkable that the existence of Nash equilibria in compact continuous games can be established in *any* topological vector space (not necessarily normed, metric, Hausdorff and/or locally convex), as shown by Theorem 4.7.2, a result due to Reny [174]. This is due to the fact that compactness allows us, by contradiction, to reduce the analysis to finite dimensions, where there is a unique Hausdorff topology (that may be derived from the Euclidean norm).

4.12 Exercises

Exercise 1. Nash equilibria computations in finite games

Compute all pure and mixed Nash equilibria of the following finite games:

(1) Two players:

	L	R
T	(6, 6)	(2, 7)
B	(7, 2)	(0, 0)

	L	R
T	(2, −2)	(−1, 1)
B	(−3, 3)	(4, −4)

	L	R
T	(1, 0)	(2, 1)
B	(1, 1)	(0, 0)

	L	M	R
T	(1, 1)	(0, 0)	(8, 0)
M	(0, 0)	(4, 4)	(0, 0)
B	(0, 8)	(0, 0)	(6, 6)

(2) Three players (player 1 chooses a row, player 2 a column, and player 3 a matrix):

	L	R
T	(1, 1, −1)	(0, 0, 0)
B	(0, 0, 0)	(0, 0, 0)

W

	L	R
T	(0, 0, 0)	(0, 0, 0)
B	(0, 0, 0)	(1, 1, −1)

E

Exercise 2. A Cournot competition

A group of n fishermen exploit a lake. If each fisher i takes a quantity $x^i \geqslant 0$, the unit price is obtained from the inverse demand as $p = \max\{1 - \sum_{i=1}^{n} x^i, 0\}$. Each fisher i sells his quantity x^i at price p and maximizes his revenue. Assume the cost of production to be zero (unrealistic but mathematically interesting).

(1) Write the normal form of the game.
(2) Compute all Nash equilibria in pure strategies.
(3) Compare with the monopoly ($n = 1$) and the perfect competition ($n \to \infty$) cases.

Exercise 3. Feasible Payoffs

Denote by j the complex number $e^{\frac{2i\pi}{3}}$, and by f the mapping from $\mathbb{C}^3$ to $\mathbb{C}$ defined by $f(a, b, c) = abc$ (the product as complex numbers). Let G be a three-player finite game where $S^1 = \{1, j\}$, $S^2 = \{j, j^2\}$, and $S^3 = \{j^2, 1\}$. For a strategy profile (a, b, c) in $S = S^1 \times S^2 \times S^3$, the payoff of player 1, $g^1(a, b, c)$, is the real part of $f(a, b, c)$, and the payoff of player 2 is the imaginary part of $f(a, b, c)$. Player 3's payoff is identically zero.

(1) Compute the payoff function $g(x, y, z)$ of the mixed extension of G.
(2) Deduce that the set of feasible payoffs in G is not contractible.

Exercise 4. Supermodular Games

(1) *Tarski's Theorem* (Tarski [208])
    Consider the Euclidean space $\mathbb{R}^n$ endowed with the partial order $x \geqslant y$ if and only if $x_k \geqslant y_k$ for all coordinates $k = 1, \ldots, n$. A subset L of $\mathbb{R}^n$ is a lattice if for all x and y in S, $x \vee y = \max\{x, y\} \in L$ and $x \wedge y = \min\{x, y\} \in L$, where $(\max\{x, y\})_k = \max(x_k, y_k)$ and similarly for the min. Let L be a compact non-empty lattice of $\mathbb{R}^n$.

    (a) Show that for any non-empty subset A of L, $\sup A \in L$ and $\inf A \in L$.
    (b) Deduce that L has a greatest and a smallest element.
    (c) Let f be a function from L to itself, monotonic with respect to the partial order. Show that f has a fixed point.

(2) *Supermodular games* (Topkis [212])
    Consider a strategic game $G = (N, (S^i)_{i \in I}, (g^i)_{i \in I})$ where for each player $i \in I$ the set of strategies S^i is a compact and non-empty lattice of $\mathbb{R}^{m_i}$ and g^i is upper semi-continuous in s^i for every $s^{-i} \in S^{-i}$ and $i \in I$. The game G is *supermodular* if:

    (i) g^i has increasing differences: for all $s^i \geqslant s'^i$ and $s^{-i} \geqslant s'^{-i}$,

    $$g^i(s^i, s^{-i}) - g^i(s'^i, s^{-i}) \geqslant g^i(s^i, s'^{-i}) - g^i(s'^i, s'^{-i}).$$

    (ii) g^i is supermodular in s^i: for all $s^{-i} \in S^{-i}$,

    $$g^i(s^i, s^{-i}) + g^i(s'^i, s^{-i}) \leqslant g^i(s^i \vee s'^i, s^{-i}) + g^i(s^i \wedge s'^i, s^{-i}).$$

    (a) Show that for all i and s^{-i}, $\mathrm{BR}^i(s^{-i})$ is a compact non-empty lattice of $\mathbb{R}^{m_i}$.
    (b) Suppose $s^{-i} \geqslant s'^{-i}$. Show that $\forall t'^i \in \mathrm{BR}^i(s'^{-i})$, $\exists t^i \in \mathrm{BR}^i(s^{-i})$ such that $t^i \geqslant t'^i$.
    (c) Deduce that G has a Nash equilibrium.

(3) *Application: Cournot Duopoly*
    Consider two firms 1 and 2 in competition to produce and sell a homogenous good. If firm $i = 1, 2$ produces the quantity $q^i \in [0, Q^i]$, it gets the payoff $g^i(q^i, q^j) = q^i P^i(q^i + q^j) - C^i(q^i)$ where the price function for firm i (the inverse demand) is P^i and its production cost is C^i. Suppose P^i and C^i are of class $\mathcal{C}^1$ and that the marginal revenue $P^i + q^i \partial P^i / \partial q^i$ is decreasing in q^j. Show the existence of a Cournot–Nash equilibrium.

Exercise 5. A Minority Game

Consider a symmetric three-player finite game where each player chooses one of two rooms and wins 1 if he is alone and zero otherwise. Compute all Nash equilibria in pure and mixed strategies.

Exercise 6. The Kakutani theorem via the existence of Nash equilibria for two-player games (McLennan and Tourki [131])

(1) Consider a class $\mathcal{G}$ of two-player games G defined by a finite action set $I = J$ and two families of points $(x_i)_{i \in I}$, $(y_i)_{i \in I}$ in $\mathbb{R}^K$. The payoff of player 1 is given by

$$f(i, j) = -\|x_i - y_j\|^2$$

and the payoff of 2 by

$$g(i, j) = \delta_{ij} \quad (1 \text{ if } i = j, 0 \text{ otherwise}).$$

G is thus a bimatrix game and has an equilibrium (σ, τ).
Prove that the support supp (τ) of τ is included in supp (σ), then that supp (σ) is included in $\{i \in I; x_i \text{ minimizes } \|x_i - z\|^2\}$, where z is the convex combination of the y_j induced by τ:

$$z = \sum_{j \in I} \tau_j y_j.$$

(2) Let C be a non-empty convex compact subset of $\mathbb{R}^K$ and F a u.s.c. correspondence with compact convex non-empty values from C to itself.
We inductively define the games $G_n = G(x_1, \ldots, x_n; y_1, \ldots, y_n)$ in $\mathcal{G}$ as follows: x_1 is arbitrary, $y_1 \in F(x_1)$; given an equilibrium (σ_n, τ_n) of G_n, introduce $x_{n+1} = \sum_{i=1}^n \tau_n(i) \, y_i$ as above and then $y_{n+1} \in F(x_{n+1})$.

  (a) Let x^* be an accumulation point of the sequence $\{x_n\}$. Let $\varepsilon > 0$ and N be such that x_{N+1} and $x_m \in B(x^*, \varepsilon)$ for some $m \leqslant N$. Consider the equilibrium (σ_N, τ_N) of $G_N = G(x_1, \ldots, x_N; y_1, \ldots, y_N)$ and show that $\{x_i, i \in S(\sigma_N)\} \subset B(x^*, 3\varepsilon)$ and $\{x_i, i \in S(\tau_N)\} \subset B(x^*, 3\varepsilon)$. Conclude that $x_{N+1} \in \text{co}\{\bigcup_z F(z); z \in B(x^*, 3\varepsilon)\}$.

  (b) Deduce the existence of a fixed point for F.

Exercise 7. Convex games [155]

A *convex game* is given by strategy sets S_i and payoff functions G_i from $S = \prod_{j \in I} S_j$ to $\mathbb{R}$, $i \in I$, satisfying:

  (i) S_i is a compact convex subset of a Euclidean space, for each $i \in I$.
  (ii) $G_i(\cdot, s_{-i})$ is concave on S_i, for all s_{-i}, for each $i \in I$.
  (iii) $\sum_{i=1}^n G_i$ is continuous on S.
  (iv) $G_i(s_i, \cdot)$ is continuous on S_{-i}, for all s_i, for each $i \in I$.

We introduce the function Φ defined on $S \times S$ by

$$\Phi(s,t) = \sum_{i=1}^{n} G_i(s_i, t_{-i}).$$

(1) Prove that t is a Nash equilibrium iff

$$\Phi(s,t) \leqslant \Phi(t,t), \quad \forall s \in S.$$

(2) Let us show the existence of an equilibrium of the convex game.
By contradiction we assume that for each $t \in S$ there exists an $s \in S$ with

$$\Phi(s,t) > \Phi(t,t).$$

(a) Show that the family

$$(O_s = \{t \in S; \Phi(s,t) > \Phi(t,t)\})_{s \in S}$$

defines an open cover of S.
(b) Deduce the existence of a finite family $(s^k)_{k \in K}$ with

$$\forall t \in S, \quad \max_{k \in K} \Phi(s^k, t) > \Phi(t,t).$$

(c) Observe then that Θ defined by

$$\Theta(t) = \frac{\sum_{k \in K} \phi_k(t) s^k}{\sum_k \phi_k(t)},$$

with $\phi_k(t) = (\Phi(s^k, t) - \Phi(t,t))^+$, is a continuous map from S to itself, hence the existence of a fixed point t^* for Θ.
(d) Finally obtain a contradiction since $\phi_k(t^*) > 0$ implies $\Phi(s^k, t^*) > \Phi(t^*, t^*)$.

4.13   Comments

The concept of Nash equilibrium (and the existence proof) first appeared in Nash's thesis. Then Nash provided several proofs for different frameworks, but all related to a fixed point argument.

Several attempts to define a similar concept had been made earlier (starting with Cournot) but it is interesting to see that the formal definition appeared during the same period as several results corresponding to the tools used (Fan, Kakutani)—as well as the related proof of existence in the economic theory of general equilibrium (Arrow, Debreu), while the related concept was introduced by Walras much earlier.

The connection between equilibria and value/optimal strategies in zero-sum games is tricky: existence of value implies existence of ε-equilibria for all $\varepsilon > 0$. However given optimal strategies (s, t) the optimality condition for player 1: $g(s, t') \geqslant g(s, t)(= v)$, $\forall t'$ corresponds to the best response (or equilibrium condition) of player 2. In terms of value one checks the payoff letting the other player's strategy vary while at equilibrium the player compares with his own set of strategies.

Note also that the interpretation of equilibria can differ in some contexts. In Nash's initial formulation the equilibrium profile s is a norm or a reference and the equilibrium condition is a property that each player satisfies given his payoff function. In particular, the fact that s is an equilibrium is not known to the players since i does not know g^{-i}, hence cannot check the equilibrium condition.

For some other authors, s is a collection of beliefs, s^{-i} being player $i's$ belief in the opponent's behavior. Then the information after the play is relevant, for example, to check if the actual moves are consistent with the beliefs. More generally, in the presence of signals this leads to the notion of conjectural or self-confirming equilibria (see some comments in [200]). The players can also hold private beliefs that are jointly inconsistent but compatible, for each player, with his own observation (Selten).

An important line of research is concerned with the efficiency of Nash equilibria (equilibrium payoffs are generically inefficient, see e.g. [49]) and a measure of inefficiency has been proposed by Koutsoupias and Papadimitriou [109] under the name "price of anarchy". It is extremely popular in computer science and has generated a large amount of literature. A connected area corresponds to mechanism design where these tools are used to generate efficient equilibrium outcomes, in particular for auctions.

McLennan and Tourky [131] proved that the existence of a mixed Nash equilibrium in every finite two-player game implies Kakutani's theorem (in finite dimensions) and so Brouwer's theorem (see Exercise 6). It has also been proved that finding a mixed equilibrium in a two-player finite game belongs to the same complexity class as finding a fixed point of a continuous function [36]. This shows that all the difficulty of fixed point theorems is captured by equilibria in two-player finite games.

Chapter 5
Equilibrium Manifolds and Dynamics

5.1 Introduction

The previous chapter was devoted to the properties of equilibria in a given game. We consider now the corresponding set as the underlying game varies in a family. We analyze structural and dynamical features that could be used to select among equilibria.

A first approach studies how the set of Nash equilibria changes as a function of the game. This allows us to check robustness with respect to payoff perturbations and to define "essential" equilibria. A second approach checks for dynamic stability. It was initiated by models of evolution in biology which introduce dynamics compatible with myopic rational behavior. We show that under some conditions, Nash equilibrium is an attractor of some "natural" dynamics.

The chapter starts by showing that equilibria of several classes of finite and infinite games may be expressed as solutions of some variational inequalities.

Then in the framework of finite games, we prove the structure result of Kohlberg and Mertens, which establishes that the projection mapping from the manifold of equilibria to the space of underlying games is homotopic to the identity map. This has several consequences. For example, it implies that, generically, the set of Nash equilibria is finite and odd, and that every game admits an essential component.

The last sections define a large class of "natural" dynamics and prove their convergence to Nash equilibria in potential games. Finally, the famous biological concept of evolutionary stable strategies [130] is defined and its link with asymptotic stability with respect to the well-known replicator dynamics is established.

© Springer Nature Switzerland AG 2019
R. Laraki et al., *Mathematical Foundations of Game Theory*, Universitext,
https://doi.org/10.1007/978-3-030-26646-2_5

5.2 Complements on Equilibria

We study here classes of games where the equilibrium characterization has a specific formulation that allows for a more precise analysis: structure of the set, associated sub-classes of games and related dynamics.

5.2.1 Equilibria and Variational Inequalities

For several classes of games, Nash equilibria can be represented as solutions of variational inequalities.

5.2.1.1 Finite Games

I is the finite set of players and for each $i \in I$, S^i is the finite set of strategies of player i and $g^i : S = \prod_j S^j \longrightarrow \mathbb{R}$ his payoff function, with the usual multilinear extension to $\Sigma = \prod_j \Sigma^j$, where $\Sigma^j = \Delta(S^j)$, the set of mixed strategies of player i, is the simplex over S^j.

Definition 5.2.1 The *vector payoff function* $Vg^i : \Sigma^{-i} \longrightarrow \mathbb{R}^{S^i}$ is defined by

$$Vg^i(\sigma^{-i}) = \{g^i(s^i, \sigma^{-i}); s^i \in S^i\}.$$

Then $g^i(\tau^i, \sigma^{-i}) = \sum_{s^i} g^i(s^i, \sigma^{-i})\tau^i(s^i) = \langle Vg^i(\sigma^{-i}), \tau^i \rangle$ and $\sigma \in \Sigma$ is a Nash equilibrium iff

$$\langle Vg^i(\sigma^{-i}), \sigma^i - \tau^i \rangle \geqslant 0, \qquad \forall \tau^i \in \Sigma^i, \forall i \in I.$$

5.2.1.2 Concave Games

I is the finite set of players and for each $i \in I$, $X^i \subset H^i$, (H^i Hilbert space), is the convex set of strategies of player i and $G^i : X = \prod_j X^j \longrightarrow \mathbb{R}$ his payoff function.

Assume G^i is concave w.r.t. x^i and of class C^1 w.r.t. x, $\forall i \in I$.

Then $x \in X$ is a Nash equilibrium iff

$$\langle \nabla_i G^i(x), x^i - y^i \rangle \geqslant 0, \qquad \forall y^i \in X^i, \forall i \in I,$$

where ∇_i stands for the gradient of G^i w.r.t. x^i, see Sect. 4.7.3. This is an extension of the above characterization for finite games because $\nabla_i G^i(\sigma) = Vg^i(\sigma^{-i})$.

5.2.1.3 Population Games

I is the finite set of populations of non-atomic agents and for each $i \in I$, S^i is the finite set of strategies of population i and $X^i = \Delta(S^i)$ is the simplex over S^i.

$x^i(s^i)$ is the proportion of agents in population i that play s^i.

Consider $K^i : S^i \times X \longrightarrow \mathbb{R}$, where $K^i(s^i, x)$ is the payoff of an agent of population i using the strategy s^i, given the configuration $x = (x^j)_{j \in I}$.

In this framework the natural notion of equilibrium is expressed as follows.

Definition 5.2.2 $x \in X$ is a *Nash/Wardrop equilibrium* [226] if

$$x^i(s^i) > 0 \Longrightarrow K^i(s^i, x) \geqslant K^i(t^i, x), \qquad \forall t^i \in S^i, \forall i \in I.$$

In fact, by changing his strategy an agent does not affect the configuration x, hence at equilibrium, any strategy chosen in population i maximizes $K^i(\cdot, x)$ over S^i.

An alternative characterization is given by the following:

Proposition 5.2.3 *$x \in X$ is a Nash/Wardrop equilibrium iff*

$$\langle K(x), x - y \rangle \geqslant 0, \qquad \forall y \in X,$$

where

$$\langle K(x), x - y \rangle = \sum_{i \in I} \langle K^i(x), x^i - y^i \rangle,$$

and

$$\langle K^i(x), x^i - y^i \rangle = \sum_{s^i \in S^i} K^i(s^i, x)[x^i(s^i) - y^i(s^i)].$$

Proof The variables being independent,

$$\langle K(x), x - y \rangle \geqslant 0, \qquad \forall y \in X,$$

is equivalent to

$$\langle K^i(x), x^i - y^i \rangle \geqslant 0, \qquad \forall y^i \in X^i, \forall i \in I,$$

and this means that all s^i in supp (x^i) maximize $K^i(\cdot, x)$. $\qquad\square$

5.2.1.4 General Evaluation

Consider a finite collection of non-empty convex compact subsets X^i of Hilbert spaces H^i, and *evaluation* mappings $\Phi^i : X \to H^i$, $i \in I$, with $X = \prod_j X^j$.

Definition 5.2.4 NE(Φ) is the set of $x \in X$ satisfying

$$\langle \Phi(x), x - y \rangle \geqslant 0, \qquad \forall y \in X, \tag{5.1}$$

where $\langle \Phi(x), x - y \rangle = \sum_i \langle \Phi^i(x), x^i - y^i \rangle_{H^i}$.

Note that all the previous sets of equilibria can be written this way.

Denote by Π_X the projection from $H = \prod_i H^i$ to the closed convex set X and by **T** the map from X to itself defined by

$$\mathbf{T}(x) = \Pi_X[x + \Phi(x)]. \tag{5.2}$$

Proposition 5.2.5 NE(Φ) *is the set of fixed points of* **T**.

Proof The characterization of the projection gives

$$\langle x + \Phi(x) - \Pi_X[x + \Phi(x)], y - \Pi_X[x + \Phi(x)] \rangle \leqslant 0, \qquad \forall y \in X,$$

hence $\Pi_X[x + \Phi(x)] = x$ is the solution iff $x \in$ NE(Φ). $\qquad\square$

Corollary 5.2.6 *Assume the evaluation* Φ *is continuous on* X. *Then* NE(Φ) $\neq \varnothing$.

Proof The map $x \mapsto \Pi_C[x + \Phi(x)]$ is continuous from the convex compact set X to itself, hence a fixed point exists (Theorem 4.11.4). $\qquad\square$

5.2.2 Potential Games

This class of games exhibits nice properties in terms of equilibria and dynamics.

5.2.2.1 Finite Games

(Monderer and Shapley [146])

Definition 5.2.7 A real function P defined on Σ is a *potential* for the game (g, Σ) if $\forall s^i, t^i \in S^i, u^{-i} \in S^{-i}, \forall i \in I$

$$g^i(s^i, u^{-i}) - g^i(t^i, u^{-i}) = P(s^i, u^{-i}) - P(t^i, u^{-i}). \tag{5.3}$$

This means that the impact due to a change of strategy of player i is the same on g^i and on P, for all $i \in I$, whatever the choice of $-i$.

5.2.2.2 Evaluation Games

We extend Sandholm's definition [184] for population games to general evaluations as defined in Sect. 5.2.1.4.

Definition 5.2.8 A real function W, of class C^1 on a neighborhood Ω of X, is a *potential* for Φ if for each $i \in I$, there is a strictly positive function $\mu^i(x)$ defined on X such that

$$\langle \nabla_i W(x) - \mu^i(x)\Phi^i(x), y^i \rangle = 0, \quad \forall x \in X, \forall y^i \in TX^i, \forall i \in I, \tag{5.4}$$

where $TX^i = \{y \in \mathbb{R}^{|S^i|}, \sum_{p \in S^i} y_p = 0\}$ is the tangent space to X^i.

A particular case is obtained when $\Phi(x)$ is the gradient of W at x. In the finite case this gives back $\nabla_i W = V g^i$.

5.3 Manifolds of Equilibria

Here we deal with finite games. Each action set S^i is finite, with cardinality m^i, $i \in I$ and $m = \Pi_i m^i$. A game g is thus identified with a point in $\mathbb{R}^{Nm}$. Consider the manifold of equilibria obtained by taking payoffs as parameters. The equilibrium equations correspond to a finite family of polynomial inequalities involving the variables (g, σ)

$$F_k(g, \sigma) \leqslant 0, \quad k \in K,$$

where $g \in R^{Nm}$ is the game and σ the strategy profile.

Let $\mathcal{G}$ be the family of games (recall that the players and their pure strategy sets are fixed) and $\mathcal{E}$ be the graph of the equilibrium correspondence

$$\mathcal{E} = \{(g, \sigma); g \in \mathcal{G}, \sigma \text{ equilibria of } g\}$$

that one extends by continuity to the compactification $\overline{\mathcal{G}}$ of $\mathcal{G}$, denoted by $\overline{\mathcal{E}}$.

Theorem 5.3.1 (Kohlberg and Mertens (1986)) *The projection π from $\overline{\mathcal{E}}$ to $\overline{\mathcal{G}}$ is homotopic to a homeomorphism.*

Proof One writes the payoff g as a function of $\tilde{g}$ and h, where for each i, h^i is a vector in $\mathbb{R}^{m^i}$ with

$$V g^i(s^{-i}) = V \tilde{g}^i(s^{-i}) + h^i$$

and $\tilde{g}$ satisfies the normalization condition

$$\sum_{s^{-i} \in S^{-i}} V \tilde{g}^i(s^{-i}) = 0.$$

Let us define, for $t \in [0, 1]$, a map Π_t from $\mathcal{E}$ to $\mathcal{G}$, where $g = (\tilde{g}; h)$, by

$$\Pi_t(g, \sigma) = (\tilde{g}; t(\sigma + Vg(\sigma)) + (1 - t)h),$$

which is the announced homotopy.

Clearly $\Pi_0 = \pi$. Let us check that $\Pi_t(\infty) = \infty, \forall t \in [0, 1]$.
Write $\| \cdot \|$ for the infinity norm, so that a neighborhood of ∞ takes the form $\{\|x\| \geqslant M\}$. Assume then that $\|(g, \sigma)\| \geqslant 2R + 1$, hence $\|g\| \geqslant 2R + 1$ so that either $\|\tilde{g}\| \geqslant R$, thus $\|\Pi_t(g, \sigma)\| \geqslant R$, or $\|h\| \geqslant 2R + 1$ and $\|\tilde{g}\| \leqslant R$. In this last case one obtains $\|t(\sigma + Vg(\sigma)) + (1 - t)h - h\| \leqslant \|(\sigma + Vg(\sigma)) - h\| \leqslant 1 + \|\tilde{g}\| \leqslant R + 1$, which implies $\|t(\sigma + Vg(\sigma)) + (1 - t)h\| \geqslant R$. $\qquad\square$

It remains to prove the next result:

Lemma 5.3.2 $\varphi = \Pi_1$ *is a homeomorphism from* $\overline{\mathcal{E}}$ *to* $\overline{\mathcal{G}}$.

Proof Define, starting from $g = (\tilde{g}, z)$, with $z = \{z^i\}$ and $z^i \in \mathbb{R}^{m^i} : \sigma = \Pi_\Sigma(z)$ and $\ell_s^i = z_s^i - \sigma_s^i - \tilde{g}^i(s, \sigma^{-i})$.

Then $\psi(g) = \psi(\tilde{g}; z) = ((\tilde{g}, \ell), \sigma)$

(i) is mapping from $\overline{\mathcal{G}}$ in $\overline{\mathcal{E}}$, and satisfies
(ii) $\varphi \circ \psi = \mathrm{Id}_{\overline{\mathcal{G}}}$;
(iii) $\psi \circ \varphi = \mathrm{Id}_{\overline{\mathcal{E}}}$.

For the first point, let us check that σ is an equilibrium in the game $\gamma = (\tilde{g}, \ell)$. Using Proposition 5.2.5, we compute $\Pi_\Sigma(\sigma + \tilde{V}g(\sigma) + \ell) = \Pi_\Sigma(z) = \sigma$, hence the result.

For (ii), starting from $g = (\tilde{g}, z)$ one obtains, via ψ, the pair $(\gamma = (\tilde{g}, \ell); \sigma)$ then through φ, the game $(\tilde{g}, \sigma + V\gamma(\sigma))$. But $V\gamma(\sigma) = \tilde{V}g(\sigma) + \ell = \tilde{V}g(\sigma) + z - \sigma - \tilde{V}g(\sigma)$, thus $\sigma + V\gamma(\sigma) = z$ and the final game is g.

Finally for (iii), starting from a pair (g, σ) in $\mathcal{E}$ with $g = (\tilde{g}, h)$, one obtains $(\tilde{g}, \sigma + Vg(\sigma))$ in $\mathcal{G}$. Thus $z = \sigma + Vg(\sigma)$, but since σ is an equilibrium in g, $\sigma = \Pi_\Sigma(\sigma + Vg(\sigma)) = \Pi_\Sigma(z)$. Hence the image of z on Σ is σ and then $\ell = z - \sigma - \tilde{V}g(\sigma) = \sigma + Vg(\sigma) - \sigma - \tilde{V}g(\sigma) = h$. $\qquad\square$

Let g be a game and NE(g) its set of equilibria, which consists of a finite family of connected components $C_k, k \in K$ (Theorem 4.9.2).

Definition 5.3.3 C_k is *essential* if for any neighborhood V of C_k in Σ, there exists a neighborhood W of g in $\mathcal{G}$ such that for any $g' \in W$ there exists a $\sigma \in$ NE$(g') \cap V$.

Proposition 5.3.4 *(i) Generically, the set of equilibria is finite and odd.*
(ii) Every game has an essential component.

Proof (i) Generically an equilibrium is isolated (by semi-algebraicity) and transverse to the projection from $\mathcal{E}$ to $\mathcal{G}$. The degree is 1 if the projection preserves the orientation and -1 if it reverses it. The global degree of the projection, which is the sum over all components, is invariant under homotopy, thus equal to 1 (due to the homeomorphism): thus there are $p + 1$ equilibria with degree $+1$ and p with degree -1, see [93].

(ii) By induction it is enough to show that if $\mathrm{NE}(g)$ is contained in $U \cup V$, where U and V are two open sets with disjoint closures, then there exists a neighborhood W of g such for any $g' \in W$, $\mathrm{NE}(g') \cap U \neq \varnothing$ or for any $g' \in W$, $\mathrm{NE}(g') \cap V \neq \varnothing$.

Let $\Psi(g)$ be the graph of the best response correspondence at g (in $\Sigma \times \Sigma$). There exists a neighborhood C of $\Psi(g)$ (with convex sections) such that the intersection with the diagonal belongs to $U \cup V \times U \cup V$.

By contradiction assume the existence of g_1 near g with $\mathrm{NE}(g_1) \cap U = \varnothing$ and similarly for g_2 and V. On the other hand one can assume that $\Psi(g_i) \subset C$. Let α be a continuous function from Σ to $[0, 1]$, with value 1 on U and 0 on V. The correspondence defined by $T(\sigma) = \alpha(\sigma)\Psi(g_1)(\sigma) + (1 - \alpha(\sigma))\Psi(g_2)(\sigma)$ is u.s.c. with convex values. Its graph is included in C, hence its non-empty set of fixed points is included in $U \cup V$. Consider such a point, $\sigma = T(\sigma)$. If $\sigma \in U$, $\alpha(\sigma) = 1$ and σ is a fixed point of $\Psi(g_1)$, hence in V. We obtain a similar contradiction if $\sigma \in V$. $\qquad \square$

As an example, consider the next family of games, with parameter $\alpha \in \mathbb{R}$:

	L	R
T	$(\alpha, 0)$	$(\alpha, 0)$
M	$(1, -1)$	$(-1, 1)$
B	$(-1, 1)$	$(1, -1)$

the equilibrium correspondence is given by (Fig. 5.1):

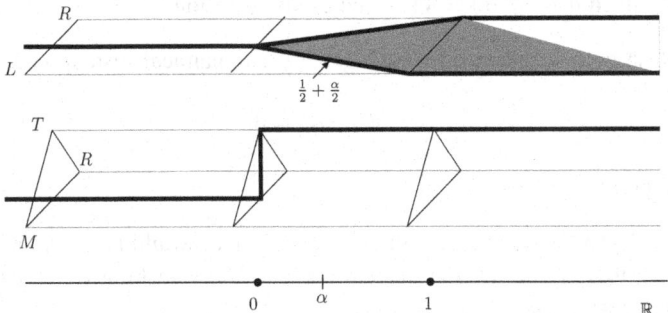

Fig. 5.1 The equilibrium correspondence

See Exercise 4.

5.4   Nash Vector Fields and Dynamics

The approach here is to consider the equilibria as the zero of a map and to define the associated dynamics.

Definition 5.4.1 A *Nash field* is a continuous map (or a u.s.c. correspondence) Ψ from $\mathcal{G} \times \Sigma$ to Σ such that

$$\mathrm{NE}(g) = \{\sigma \in \Sigma; \Psi(g, \sigma) = \sigma\}, \forall g \in \mathcal{G}.$$

If no continuity w.r.t. g is asked for, one can for example select $\sigma^g \in \mathrm{NE}(g)$ for each g and let $\Psi(g, \sigma) = d(\sigma)\sigma^g + [1 - d(\sigma)]\sigma$, where $d(\sigma)$ is the minimum of 1 and the distance from σ to $\mathrm{NE}(g)$.

Proposition 5.4.2 *The following maps are Nash fields:*

1. *(Nash [151])*

$$\Psi(g, \sigma)^i(s^i) = \frac{\sigma^i(s^i) + (g^i(s^i, \sigma^{-i}) - g^i(\sigma))^+}{1 + \Sigma_{t^i}(g^i(t^i, \sigma^{-i}) - g^i(\sigma))^+};$$

2. *(Gul, Pearce and Stacchetti [87]) (recall that Π_Σ is the projection on the convex compact set Σ)*

$$\Psi(g, \sigma) = \Pi_\Sigma(\{\sigma^i + V g^i(\sigma^{-i})\}).$$

Proof The continuity is clear in both cases.
   (1) Proof of Theorem 4.6.2.
   (2) For a fixed g, $\Psi(g, \sigma) = \mathbf{T}(\sigma)$ and use Proposition 5.2.5.   □

Each Nash field Ψ induces, for each game g, a dynamical system on Σ,

$$\dot{\sigma} = \Psi(g, \sigma) - \sigma,$$

whose rest points are $\mathrm{NE}(g)$.

Remark 5.4.3 An alternative definition of a Nash field would be a continuous map from $\mathcal{G} \times \Sigma$ to T_Σ (the tangent vector space at Σ), vanishing on $\mathcal{E}$ and inward pointing at the boundary of Σ.

Remark 5.4.4 Every component of the set of fixed points has an index and the sum of indices is 1, which is the Euler characteristic of the simplex Σ (the Poincaré–Hopf Theorem, see Milnor [143, p. 35]). Moreover, the index of a component C is independent of the Nash field and is equal to the local degree at C of the projection π from $\mathcal{E}$ to $\mathcal{G}$ [43, 83].

For the next game, with parameter $t \in \mathbb{R}$:

	L	R
T	(t, t)	$(0, 0)$
B	$(0, 0)$	$(1 - t, 1 - t)$

one obtains for the manifold and the dynamics the following situation (Fig. 5.2):

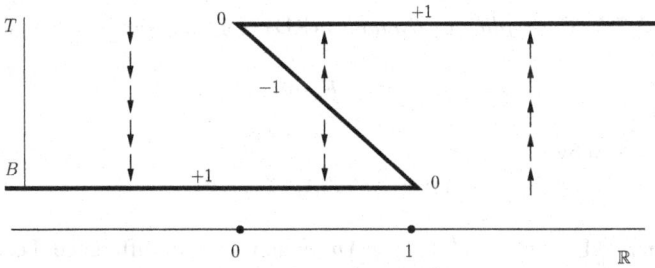

Fig. 5.2 A Nash dynamics

For $t \notin [0, 1]$ there is only one equilibrium, hence there is only one configuration for the Nash field (inwards at the boundary and vanishing only on the manifold of equilibria).

By continuity we obtain the configuration for $t \in (0, 1)$, thus the mixed equilibria has index -1: the projection reverses the orientation of the manifold—or the vector field has index -1 there.

For further results, see [45, 84, 175, 217].

5.5 Equilibria and Evolution

Here we introduce several dynamical approaches to equilibria.

5.5.1 Replicator Dynamics

Consider a symmetric two-player game defined by a matrix A in $\mathbb{R}^{K \times K}$. A_{kj} is the "fitness" (reproduction rate) of $k \in K$ within an interaction (k, j). (The payoff of player 2 is thus $B = {}^t A$.)

The first approach corresponds to the study of a single polymorphic non-atomic population, playing in pure strategies. Its composition is p if a proportion p^k of its members is of "type k". An alternative approach is described in Sect. 5.5.6.

An interaction corresponds to the matching of two agents chosen at random in the population. An agent will be of type k with probability p^k and he will face a type j with probability p^j, hence his expected fitness will be $e^k A p$.

Definition 5.5.1 A population $p \in \Delta(K)$ is *stationary* if

$$p^k > 0 \Longrightarrow e^k Ap = pAp.$$

Thus all types actually present in the population have the same reproduction rate, hence the composition of the population remains the same after an interaction.

If p is stationary with full support, it is a symmetric equilibrium; on the other hand each pure strategy is stationary.

Definition 5.5.2 The *replicator dynamics* (RD) (for one population) is defined on $\Delta(K)$ by

$$\dot{p}_t = F(p_t),$$

where F is given by

$$F^k(p) = p^k(e^k Ap - pAp).$$

Note that $\frac{d}{dt}\text{Log}(p_t^k) = e^k Ap - pAp$ which is the difference between the (expected) fitness of k and the average in the population. This dynamics preserves the simplex since $\sum_k F^k(p) = pAp - \sum_k p^k pAp = 0$. One could also define a stationary population as a rest point of the replicator dynamics.

5.5.2 The RSP Game

Consider the following game ("rock, scissors, paper"):

$(0, 0)$	$(a, -b)$	$(-b, a)$
$(-b, a)$	$(0, 0)$	$(a, -b)$
$(a, -b)$	$(-b, a)$	$(0, 0)$

where a and b are two parameters > 0.

Proposition 5.5.3 *The only Nash equilibrium is* $E = (1/3, 1/3, 1/3)$. *It is an attractor for the replicator dynamics if* $a > b$ *and a repulsor if* $a < b$.

Proof Let us compute the average fitness for $p \in \Delta(K)$:

$$pAp = (a - b)(p^1 p^2 + p^2 p^3 + p^1 p^3)$$
$$= \frac{(a - b)}{2}(1 - \|p\|^2).$$

Let $V(p) = \Pi_{k=1}^3 p^k$, which is maximal at E. Since: $\dot{p}_t^1 = p_t^1(ap_t^2 - bp_t^3 - p_t Ap_t)$, we obtain

Fig. 5.3 Case $a > b$ and $a < b$

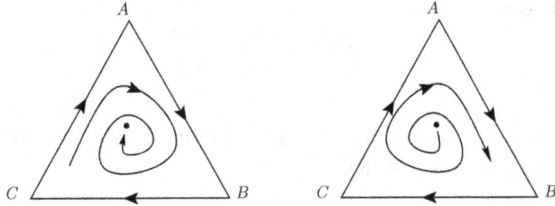

$$\frac{d}{dt} \log V(p_t) = \sum_{k=1}^{3} \frac{\dot{p}_t^k}{p_t^k} = (a - b) - 3 p_t A p_t = \frac{(a - b)}{2}(3\|p_t\|^2 - 1).$$

Thus for $a > b$, V increases as long as $\|p_t\|^2 \geqslant 1/3$, which implies convergence to E. Similarly E is a repulsor for $a < b$ (Fig. 5.3). $\qquad\square$

5.5.3 Potential Games

In the general case of a finite game (g, Σ), the replicator dynamics takes the following form, for each player $i \in I$ and each of his actions $s \in S^i$:

$$\dot{\sigma}_t^{is} = \sigma_t^{is}[g^i(s, \sigma_t^{-i}) - g^i(\sigma_t)].$$

In particular, for a potential game (the partnership game $(A = {}^t A = B)$ is an example of a potential game for two players), the replicator dynamics can be expressed in terms of the potential P:

$$\dot{\sigma}_t^{is} = \sigma_t^{is}[P(s, \sigma_t^{-i}) - P(\sigma_t)].$$

Proposition 5.5.4 *In a potential game, the potential is a strict Lyapounov function for the replicator dynamics and the set of stationary populations.*

Proof Let $f_t = P(\sigma_t)$. Thus $\dot{f}_t = \sum_i P(\dot{\sigma}_t^i, \sigma_t^{-i})$ by linearity, but one has

$$P(\dot{\sigma}_t^i, \sigma_t^{-i}) = \sum_s \dot{\sigma}_t^{is} P(s, \sigma_t^{-i}) = \sum_s \sigma_t^{is}[P(s, \sigma_t^{-i}) - P(\sigma_t)] P(s, \sigma_t^{-i}).$$

Let us add

$$0 = \sum_s \sigma_t^{is}[P(s, \sigma_t^{-i}) - P(\sigma_t)] P(\sigma_t)$$

to get

$$P(\dot{\sigma}_t^i, \sigma_t^{-i}) = \sum_s \sigma_t^{is}[P(s, \sigma_t^{-i}) - P(\sigma_t)]^2.$$

Thus $\dot{f}_t$ is a sum of non-negative terms and the minimum, which is 0, is obtained on the rest points of (RD). □

An important example corresponds to *congestion games*, see e.g. Chapter 18 in Nisan et al. [156] and Exercise 3.

5.5.4 Other Dynamics

The general form of a dynamics describing the evolution of the strategic interaction in a game with evaluation $\Gamma(\Phi)$ as defined in Sect. 5.2.1.4, with $X^i = \Sigma^i = \Delta(S^i)$, S^i finite, is

$$\dot{\sigma} = \mathcal{B}_\Phi(\sigma), \quad \sigma \in \Sigma,$$

where Σ is invariant, so that for each $i \in I$, $\mathcal{B}_\Phi^i(\Sigma) \in T\Sigma^i$, which is the tangent space: $T\Sigma^i = \{t^s; \sum_{s \in S^i} t^s = 0\}$.

Here are few example of dynamics expressed in terms of the evaluation Φ.

5.5.4.1 Replicator Dynamics

(Taylor and Jonker [209])

$$\dot{\sigma}^{ip} = \sigma^{ip}[\Phi^{ip}(\sigma) - \overline{\Phi}^i(\sigma)], \quad p \in S^i, i \in I,$$

where

$$\overline{\Phi}^i(\sigma) = \langle \sigma^i, \Phi^i(\sigma) \rangle = \sum_{p \in S^i} \sigma_p^i \Phi_p^i(\sigma)$$

is the average evaluation for participant i.

5.5.4.2 Brown–von Neumann–Nash Dynamics

(Brown and von Neumann [34], Hofbauer [99])

$$\dot{\sigma}^{ip} = \hat{\phi}^{ip} - \sigma^{ip} \sum_{q \in S^i} \hat{\phi}^{iq}, \quad p \in S^i, i \in I,$$

where $\hat{\Phi}^{iq} = [\Phi^{iq}(\sigma) - \overline{\Phi}^i(\sigma)]^+$ is called the "excess evaluation" of p. (Recall that $t^+ = \max\{t, 0\}$.)

5.5.4.3 Smith Dynamics

(Smith [193])

$$\dot{\sigma}^{ip} = \sum_{q \in S^i} \sigma^{ip}[\Phi^{ip}(\sigma) - \Phi^{iq}(\sigma)]^+ - \sigma^{ip} \sum_{q \in S^i} [\Phi^{iq}(\sigma) - \Phi^{ip}(\sigma)]^+, \quad p \in S^i, i \in I,$$

where $[\Phi^i_p(\sigma) - \Phi^i_q(\sigma)]^+$ corresponds to "pairwise comparison", Sandholm [185].

5.5.4.4 Best Response Dynamics

(Gilboa and Matsui [78])

$$\dot{\sigma}^i \in BR^i(\sigma) - \sigma^i, \quad i \in I,$$

where:

$$BR^i(\sigma) = \{y^i \in \Sigma^i, \langle y^i - z^i, \Phi^i(\sigma) \rangle \geqslant 0, \forall z^i \in \Sigma^i\}.$$

5.5.5 A General Property

It is natural to require some consistency between the objective and the dynamics.

Definition 5.5.5 The dynamics $\mathcal{B}_\Phi$ satisfies *positive correlation* (PC) if

$$\langle \mathcal{B}^i_\Phi(\sigma), \Phi^i(\sigma) \rangle > 0, \quad \forall i \in I, \forall \sigma \in \Sigma \text{ s.t. } \mathcal{B}^i_\Phi(\sigma) \neq 0.$$

This corresponds to MAD (myopic adjustment dynamics, [207]): assuming the configuration given, a unilateral change along the dynamics should increase the evaluation Φ. In a discrete time framework this reads as

$$\langle \sigma^i_{n+1} - \sigma^i_n, \Phi^i(\sigma_n) \rangle \geqslant 0.$$

Proposition 5.5.6 *Consider a potential game* $\Gamma(\Phi)$ *with potential function* W. *If the dynamics* $\dot{\sigma} = \mathcal{B}_\Phi(\sigma)$ *satisfies (PC), then* W *is a strict Lyapunov function for* $\mathcal{B}_\Phi$.

Proof Let $\{\sigma_t\}_{t\geqslant 0}$ be the trajectory of $\mathcal{B}_\Phi$ and $V_t = W(\sigma_t)$ for $t \geqslant 0$. Then

$$\dot{V}_t = \langle \nabla W(\sigma_t), \dot{\sigma}_t \rangle = \sum_{i\in I} \langle \nabla^i W(\sigma_t), \dot{\sigma}_t^i \rangle = \sum_{i\in I} \mu^i(\sigma) \langle \Phi^i(\sigma_t), \dot{\sigma}_t^i \rangle \geqslant 0.$$

(Recall that $\dot{\sigma}_t \in T\Sigma$.) Moreover, $\langle \Phi^i(\sigma_t), \dot{\sigma}_t^i \rangle = 0$ holds for all i if and only if $\dot{\sigma}_t = \mathcal{B}_\Phi(\sigma_t) = 0$. $\qquad\square$

5.5.6 ESS

Returning to the framework of Sect. 5.5.1, the concept of an "Evolutionary Stable Strategy" (ESS), due to Maynard Smith [130], corresponds to the study of an asexual homogeneous population which holds a "mixed type" $p \in \Delta(K)$ and one considers its local stability.

Definition 5.5.7 $p \in \Delta(I)$ is an *ESS* if it is robust under perturbations in the sense that for each alternative $q \in \Delta(I), q \neq p$, there exists an $\varepsilon(q) > 0$ such that $0 < \varepsilon \leqslant \varepsilon(q)$ implies

$$pA((1-\varepsilon)p + \varepsilon q) > qA((1-\varepsilon)p + \varepsilon q).$$

$\varepsilon(q)$ is the threshold associated to q.

This inequality can be decomposed into

$$pAp \geqslant qAp,$$

thus p is a symmetric equilibrium, and if equality holds:

$$pAq > qAq.$$

A classical example of an ESS is a strict equilibrium.

Proposition 5.5.8 *p is an ESS iff one of the following conditions is fulfilled:*

(1) There exists an ε_0 (independent of q) such that

$$pA((1-\varepsilon)p + \varepsilon q) > qA((1-\varepsilon)p + \varepsilon q)$$

for all $\varepsilon \in (0, \varepsilon_0)$ and all $q \in \Delta(K)$, $q \neq p$.
(2) There exists a neighborhood $V(p)$ of p such that

$$pAq > qAq, \quad \forall q \in V(p), q \neq p.$$

Proof Consider the subset of the boundary ∂X of the simplex $X = \Delta(K)$ facing p, namely $\partial X_p = \{q; q \in \partial X$ with $q^k = 0$ and $p^k > 0$ for at least one component $k\}$. This defines a compact subset disjoint from p where the threshold $\varepsilon(q)$ is bounded below by some $\varepsilon_0 > 0$.

Now each $r \neq p$ in X can be written as $r = tq + (1-t)p$ with $t > 0$ and $q \in \partial X_p$, hence $(1-\varepsilon)p + \varepsilon r = (1-\varepsilon')p + \varepsilon'q$ with $\varepsilon' \leq \varepsilon$, so that as soon as $\varepsilon \leq \varepsilon_0$,

$$pA((1-\varepsilon')p + \varepsilon'q) > qA((1-\varepsilon')p + \varepsilon'q).$$

By multiplying by ε' and adding $(1-\varepsilon')pA((1-\varepsilon')p + \varepsilon'q)$ one also obtains

$$pA((1-\varepsilon')p + \varepsilon'q) > ((1-\varepsilon')p + \varepsilon'q)A((1-\varepsilon')p + \varepsilon'q),$$

which implies

$$pA((1-\varepsilon)p + \varepsilon r) > rA((1-\varepsilon)p + \varepsilon r).$$

To prove (1) we just need to show that on ∂X_p, for $\varepsilon \leq \varepsilon_0$,

$$pA((1-\varepsilon)p + \varepsilon q) > qA((1-\varepsilon)p + \varepsilon q),$$

thus one obtains for $\varepsilon \in (0, \varepsilon_0)$:

$$pA((1-\varepsilon)p + \varepsilon q) > ((1-\varepsilon)p + \varepsilon q)A((1-\varepsilon)p + \varepsilon q).$$

It remains to observe that as q varies in ∂X_p and $\varepsilon \in (0, \varepsilon_0)$ the set $((1-\varepsilon)p + \varepsilon q)$ describes a pointed neighborhood $V(p) \setminus \{p\}$.

Conversely, starting from $q \neq p$, $p' = (1-\varepsilon)p + \varepsilon q$ is in $V(p) \setminus \{p\}$ for $\varepsilon > 0$ small enough, and then $pAp' > p'Ap' = [(1-\varepsilon)p + \varepsilon q]Ap'$, from which we deduce that

$$pAp' > qAp'. \qquad \square$$

In the game RSP the only equilibrium is ESS iff $a > b$ and there is no ESS for $a < b$: in fact each pure strategy is as good as $(1/3, 1/3, 1/3)$ facing $(1/3, 1/3, 1/3)$ and gives 0 against itself while $(1/3, 1/3, 1/3)$ induces $(a-b)/3 < 0$.

Proposition 5.5.9 p *is an ESS iff* $V(x) = \prod_i(x^i)^{p^i}$ *is locally a strict Lyapounov function for the replicator dynamics.*

Proof V has a unique maximum in $\Delta(I)$ taken at p: in fact Jensen's inequality applied to $\log V$ gives

$$\sum_i p^i \log(x^i/p^i) \leq \log \sum_i x^i = 0.$$

Let $v_t = \log V(x_t)$. Then, in a neighborhood of p, one obtains

$$\dot{v}_t = \sum_i p^i \frac{\dot{x}_t^i}{x_t^i} = \sum_i p^i [e_i A x_t - x_t A x_t]$$
$$= p A x_t - x_t A x_t > 0,$$

by the previous Proposition 5.5.8.   □

For a full study of evolution games, see Hammerstein and Selten [88], Hofbauer and Sigmund [102], van Damme [214], Weibull [227] and Sandholm [185].

5.6 Exercises

Exercise 1. Potential games

(1) Let $G = (g, \Sigma)$ be a finite game with potential P (Definition 5.2.7). Show that the equilibria are the same in the game where each player's payoff is P. Deduce that G has an equilibrium in pure strategies.

(2) Example 1. Show that the following game is a potential game and compute its set of equilibria:

	b_1	b_2
a_1	$(2, 2)$	$(0, 0)$
a_2	$(0, 0)$	$(1, 1)$

(3) Example 2. Show that the prisoners dilemma is a potential game:

	b_1	b_2
a_1	$(0, 4)$	$(3, 3)$
a_2	$(1, 1)$	$(4, 0)$

with potential:

	b_1	b_2
a_1	1	0
a_2	2	1

(4) Example 3. Consider a *congestion game*: two towns are connected via a set K of roads. $u^k(t)$ denotes the payoff of each of the users of road k, if their number is t.

The game is thus defined by $S^i = K$ for each $i = 1, \ldots, n$ (each user i can choose a road s^i in K) and if $s = (s^1, \ldots, s^n)$ stands for the profile of choices, the payoff of player i choosing $s^i = k$ is $g^i(s) = g^i(k, s^{-i}) = u^k(t^k(s))$, where $t^k(s)$ is the number of players j for which $s^j = k$.

Show that G is a potential game with potential P given by

$$P(s) = \sum_{k \in K} \sum_{r=1}^{t^k(s)} u^k(r).$$

(5) Example 4. Consider a non-atomic congestion game: there are I non-atomic populations of size one and x^{ik} is the proportion of agents in population i using road k. Let $z^k = \sum_{i \in I} x^{ik}$ be the congestion on road k. Define the real function W on X by

$$W(x) = \sum_{k \in K} \int_0^{z^k} u^k(r) dr.$$

Show that W is a potential for the game.

(6) Consider a finite game with potential P. Prove that P is a Lyapounov function for the best response dynamics. Deduce that all accumulation points are Nash equilibria.

(7) Show that the dynamics defined in Sect. 5.5.4 satisfy (PC). Deduce that the set of accumulation points for the dynamics 5.5.4.2–5.5.4.4 is included in the set of equilibria.

Exercise 2. Dissipative games

Consider the evaluation framework of Sect. 5.2.1.4 (compare with the monotone games defined in Sect. 4.7.3).

Definition 5.6.1 An evaluative game $\Gamma(\Phi)$ is called *dissipative* if Φ satisfies

$$\langle \Phi(x) - \Phi(y), x - y \rangle \leqslant 0, \qquad \forall\, (x, y) \in X \times X.$$

Notice that if Φ is dissipative and derives from a potential W, W is concave. Introduce $\text{SNE}(\Phi)$ as the set of $x \in X$ satisfying:

$$\langle \Phi(y), x - y \rangle \geqslant 0, \qquad \forall y \in X. \tag{5.5}$$

(1) Prove that if Φ is continuous, then

$$\text{SNE}(\Phi) \subset NE(\Phi).$$

(2) Show that if $\Gamma(\Phi)$ is dissipative, then

$$NE(\Phi) \subset \text{SNE}(\Phi).$$

(3) Deduce that if $\Gamma(\Phi)$ is dissipative and Φ is continuous, then $NE(\Phi)$ is convex.

Exercise 3. Equilibrium correspondence
Consider a family of two-player finite games with parameter $\alpha \in \mathbb{R}$:

	G	D
H	$\alpha, 0$	$\alpha, 0$
M	$1, -1$	$-1, 1$
B	$-1, 1$	$1, -1$

Compute the set of Nash equilibria as a function of α.

Exercise 4. Fictitious play in non-zero-sum games
Consider a two-person finite game $F : S^1 \times S^2 \to \mathbb{R}^2$ and the fictitious play process defined in Sect. 2.7.

(1) Consider the anticipated payoff at stage n: $E_n^i = F^i(s_n^i, \bar{s}_{n-1}^{-i}))$ and the average payoff up to stage n (excluded) ($A_n^i = \frac{1}{n-1}\sum_{p=1}^{n-1} F^i(s_p))$. Show that

$$E_n^i \geqslant A_n^i.$$

(Monderer, Samet and Sela [144].)
Remark: This is a unilateral property: no hypothesis is made on the behavior of player $-i$.

(2) Prove that
$$F^i(s_n^i, s_{n-1}^{-i}) \geqslant F^i(s_{n-1}).$$

(The improvement principle, Monderer and Sela [145].)

(3) Consider the following two-player game, due to Shapley [189]:

$(0, 0)$	(a, b)	(b, a)
(b, a)	$(0, 0)$	(a, b)
(a, b)	(b, a)	$(0, 0)$

with $a > b > 0$. Note that the only equilibrium is $(1/3, 1/3, 1/3)$.
Prove that starting from a Pareto entry, fictitious play does not converge.

5.7 Comments

The representation of Nash equilibria as variational inequalities played an important rôle in Operations Research [39, 51], and was used in the transportation literature to obtain approximation results.

The dynamic approach has been extremely productive in evolution and biology but also shows a difference between strategic and dynamic stability requirements (see also Chap. 6).

Some concepts may lack a general existence property but have very strong dynamic properties (ESS).

On the other hand attractors may be a quite natural issue for dynamics, without necessarily consisting of a collection of fixed points.

In summary, the dynamic approach is quite satisfactory for specific classes of games (potential, dissipative, super-modular) but in general the results differ from the 0-sum case (see Fictitious Play), which exhibits some kind of "periodic regularity".

Among the open problems, we would like to prove convergence of the trajectories and not only identify the accumulation points as equilibria. A first step would be to consider smooth data in the spirit of [29].

Finally, a lot of recent results explicitly describe the set of equilibria of finite games and indicates that essentially any semi-algebraic set can be achieved at Nash equilibrium (in the space of payoffs or strategies).

Chapter 6
Games in Extensive Form

6.1 Introduction

The previous chapters dealt with games in strategic or "normal" form. This one studies games in "extensive" form. The model describes precisely the rules of the game: which player starts, what actions he can take, who plays next, what is the information before choosing an action, and what payoff is obtained at the end.

We start with the simplest and oldest model of perfect information (PI) games. Here, players successively choose actions according to some rules and at every decision node, all previous actions are observed by all the players (like in Chess). We prove Zermelo's famous theorem stating that every finite zero-sum PI game has a value in pure strategies (e.g. is *determined*) and show that optimal strategies can be computed using backward induction. This algorithm easily extends to n-player PI games, leading to the first and famous refinement of Nash equilibrium: Selten's subgame perfection. We also explore infinite PI games and provide conditions under which they are determined.

Next, we define general extensive form (EF) games. In this model, players can have imperfect and asymmetric knowledge about past actions (like in poker). In this class, for the value or an equilibrium to exist, players may need to randomize. We define several randomized strategies (mixed, behavioral, general) and show their outcome-equivalence in EF games with perfect recall, i.e. in which each player remembers what he did or learned in the past at every node where he has to play. This is the celebrated Kuhn's theorem.

Finally, after describing several nested equilibrium refinement concepts in extensive form games (the finest being the sequential equilibrium of Kreps and Wilson), links with normal form equilibrium refinements are established (in particular with Myerson's proper equilibrium). More precisely, it is shown that a proper equilibrium of a normal form game induces a sequential equilibrium in any extensive form having that normal form (note that several extensive form games may have the same normal form). The theory that studies the invariance of a Nash equilibrium under the exten-

© Springer Nature Switzerland AG 2019
R. Laraki et al., *Mathematical Foundations of Game Theory*, Universitext,
https://doi.org/10.1007/978-3-030-26646-2_6

sive form representation and other properties is called strategic stability (initiated by Kohlberg and Mertens). Some of its principles, such as the forward induction reasoning, are discussed in the last section.

6.2 Extensive Form Games with Perfect Information

A finite game with perfect information is the first and simplest model in game theory. Examples are games like Chess or Go. Players play alternately and at their turn, each player knows all past actions. The payoff, obtained when the game ends, depends on the whole sequence of actions. The model has applications in a number of fields ranging from set theory to logic, computer science, artificial intelligence, and economics.

6.2.1 Description

A finite game in *extensive form and perfect information* G is described using a tree (a finite connected directed graph without cycles). More formally, the game is defined by:

1. A finite non-empty set of *players* I;
2. A finite set of *nodes* Z;
3. An *origin* $\theta \in Z$;
4. A *predecessor* mapping $\phi : Z \setminus \{\theta\} \to Z$ satisfying: for every $z \in Z$, there is an integer l such that $\phi^l(z) = \theta$;
5. A set of *terminal* nodes (outcomes) $R := Z \setminus \text{Im}(\phi)$;
6. A set of *positions* $P := Z \setminus R = \text{Im}(\phi)$;
7. A set of *successors* for each position $p \in P$, $S(p) := \phi^{-1}(p)$;
8. A *partition* $\{P^i, i \in I\}$ of P;
9. A *payoff* $g^i(r)$ for each player $i \in I$ and each outcome $r \in R$.

The game is played as follows:

(a) Let i be the player in I such that $\theta \in P^i$.
(b) At stage $t = 1$, player i chooses a successor $p_1 \in S(\theta)$.
(c) Inductively at stage t, player j with $p_t \in P^j$ chooses a successor p_{t+1} of p_t.
(d) The game terminates when a terminal node r in R is reached.
(e) Each player $i \in I$ obtains the payoff $g^i(r)$.

To each position p is associated a *history* (the sequence of predecessors of p) and a *subgame* $G[p]$ consisting of the sub-tree generated by all successors of p. Thus, $G[p]$ is itself an extensive form game with perfect information having p as origin.

In Game 1 of Fig. 6.1 player 1 starts. He has two actions. If he chooses the right action, node p is reached, in which player 2 has 3 actions a, b and c. If he chooses a, we reach the terminal node $x \in R$. As in this example, it is usual to attribute names to actions (i.e. to edges that connect two successive nodes).

6.2.2 Strategy and Normal Form

A strategy σ^i of player i is a function on P^i which associates to each position $p \in P^i$ a successor in $S(p)$. A profile of strategies $\sigma = (\sigma^1, \ldots, \sigma^N)$ induces a unique outcome r in R.

For example, in Game 2 (Fig. 6.2) player 1 has two strategies $S^1 = \{\alpha, \beta\}$. Player 2 has $3 \times 2 = 6$ strategies: $S^2 = \{aA, aB, bA, bB, cA, cB\}$. The strategy bA means that player 2 will play the move b at node p and the move A at node q.

Definition 6.2.1 The application F that associates to each strategy profile σ the induced outcome in R is called the *normal* or *strategic form reduction*.

The normal form of game 2 is:

	aA	aB	bA	bB	cA	cB
α	x	x	y	y	z	z
β	u	v	u	v	u	v

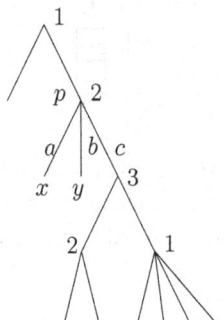

Fig. 6.1 Game 1 in extensive form and with perfect information

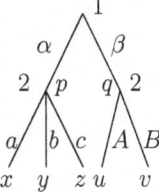

Fig. 6.2 Game 2 in extensive form and with perfect information

To complete the definition of the game, we need to specify for each element r in R a payoff $g^i(r)$ for each player $i \in I$. For example, if $g^1(x) = +1$ and $g^2(x) = -5$, this means that if terminal node x is reached, player 1 gets +1 and player 2 gets −5.

6.2.3 The Semi-reduced Normal Form

In the following game, player 1 controls two decision nodes where, in each, he has two actions. Thus, he has 4 strategies: $S^1 = \{A\alpha, A\beta, B\alpha, B\beta\}$. Player 2 has only one decision node and two actions, thus $S^2 = \{a, b\}$ (Fig. 6.3).

The *normal form* of Γ represented in the outcome space is:

	a	b
$A\alpha$	x	x
$A\beta$	x	x
$B\alpha$	y	z
$B\beta$	y	w

Note that strategies $A\alpha$ and $A\beta$ of player 1 are equivalent in a very robust sense: for every strategy of player 2, the outcome is the same. In fact, in the extensive form description, those strategies differ only on positions that they both exclude. The *semi-reduced* normal form is obtained by identifying those equivalent strategies:

	a	b
A	x	x
$B\alpha$	y	z
$B\beta$	y	w

More generally, two strategies s^i and t^i of player i are *i-payoff-equivalent* if for all $s^{-i} \in S^{-i}$, $g^i(s^i, s^{-i}) = g^i(t^i, s^{-i})$ and are *payoff-equivalent* if for all $s^{-i} \in S^{-i}$, and all $j \in I$ $g^j(s^i, s^{-i}) = g^j(t^i, s^{-i})$. Above, $A\alpha$ and $A\beta$ are payoff-equivalent, and so, are identified.

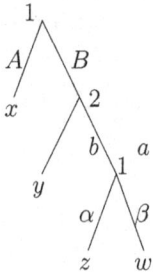

Fig. 6.3 A perfect information game Γ

6.2.4 Determinacy of Perfect Information Finite Games

A two-player game with perfect information is called *simple* if there is a partition (R^1, R^2) of R such that if the outcome is in R^i, player i wins and player $-i$ loses. Say that player i has a *winning strategy* σ^i if he can force the outcome to be in R^i:

$$\exists \sigma^i, \quad \forall \sigma^{-i} \quad F[\sigma] \in R^i.$$

That is, there is a strategy of player i such that for every strategy of the opponent, an outcome in R^i is reached. Since R^1 and R^2 are disjoint, both players cannot have a winning strategy.

Definition 6.2.2 A game is *determined* if one of the players has a winning strategy.

Theorem 6.2.3 ([231]) *Every simple finite game with perfect information is determined.*

Proof The proof is by induction on the length n of the tree. Use the convention that a determined game has value $+1$ if player 1 has a winning strategy and -1 if player 2 has a winning strategy. A game of length 1 is a one player game in which once the player plays, the game is over. If that player has a winning action, he plays it. Otherwise the other player, without playing, wins.

Suppose now that any perfect information game of length less than or equal to n has a value (i.e. is determined) and let us prove that any game of length $n + 1$ is determined.

Proof 1 (*by forward induction*) Any successor of the origin induces a subgame of duration less than or equal to n. By hypothesis, these are determined. If player 1 starts the game, it suffices for him to choose a subgame with the highest value. If this value is $+1$, he wins, otherwise, whatever he plays, player 2 has a winning strategy. If player 2 starts, the situation is similar.

Proof 2 (*by backward induction*) Positions that are predecessors of terminal nodes are one player games with duration length 1. Replace each of them by its value. The new game has a strictly smaller length and so by induction, it is determined. If a player is winning in the new game, he is also winning in the original game. Just follow the winning strategy of the new game and finish by an optimal move at nodes where all successors are terminal. □

Suppose now that the set of outcomes R contains more that two outcomes $R = \{r_1 \succ_1 r_2 \cdots \succ_1 r_n\}$ ordered with respect to the preferences of player 1 (we identify outcomes where player 1 is indifferent). A two-player game is called *strictly competitive* if player 2 has the reverse preference order over R: $r_1 \prec_2 r_2 \cdots \prec_2 r_n$. The game is *determined* if there is a k such that player 1 has a strategy that guarantees an outcome in $\{r_1, r_2, \ldots, r_k\}$ and player 2 has a strategy that guarantees an outcome in $\{r_k, r_{k+1}, \ldots, r_n\}$.

If the outcomes are interpreted as the payoff of player 1, and if strictly competitive means that the game is zero-sum, being determined is equivalent to the game having a value and players having optimal pure strategies.

Corollary 6.2.4 *Every finite strictly competitive game with perfect information is determined.*

Proof Define $R_m = \{r_1, \ldots, r_m\}$ for each m and suppose $R_0 = \varnothing$. Let $R_k = \{r_1, \ldots, r_k\}$ be the smallest set R_m, $m = 1, \ldots, n$, that player 1 can guarantee. Because player 1 cannot guarantee $R_{k-1} = \{r_1, \ldots, r_{k-1}\}$, by Zermelo's theorem, player 2 can guarantee its complement $\{r_k, r_{k+1}, \ldots, r_m\}$.   □

Chess is a finite game with perfect information where only three outcomes are possible, consequently, either one of the players has a winning strategy or both can guarantee a draw. Because of the complexity of the game, its value is not known. In practice, there are very performant computer programs which can often beat any human player. They give a score to every position and solve the game using backward induction by looking several steps ahead. For the game of Go, computer programs have recently started to beat humans. The algorithms combine deep learning and Monte Carlo search.

6.2.5 Nature as a Player

In some perfect information games, such as backgammon, some transitions are random. It is easy to extend the precedent model to that case: just add a new player (called the hazard player, or Nature), specify nodes where he plays and provide in the description of the game the probability distributions according to which Nature selects the successors for each of its decision nodes. For example, in the game of Fig. 6.4, if player 1 chooses move b, Nature can choose at random either c or d. With probability $\frac{1}{3}$ the game reaches node p_2 where player 1 can play again, and with probability $\frac{2}{3}$ the game reaches node p_3 where player 2 has to play. The strategy profile (bA, β) induces the random outcome $\frac{1}{3}y + \frac{2}{3}v$.

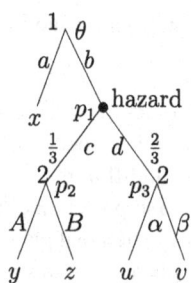

Fig. 6.4 A perfect information game with Nature

Proposition 6.2.5 *Every finite zero-sum perfect information game (with or without Nature) has a value and players have pure optimal strategies.*

Proof Let us prove the result by forward induction. If Nature starts, then for each (randomly) chosen move $k \in K$ with probability p_k, a subgame G_k is reached. By induction, each G_k has a value v_k and players have pure optimal strategies (s_k^1, s_k^2). Consequently, the original game has value $v = \sum_{k \in K} p_k v_k$, and pure optimal strategies $(s^1 = (s_k^1)_{k \in K}$ for player 1 and $s^2 = (s_k^2)_{k \in K}$ for player 2).

If player 1 starts the game, by induction, each subgame G_k has a value v_k and optimal pure strategies (s_k^1, s_k^2). Consequently, the value of the original game is $\max_k v_k$. To guarantee this value, player 1 plays a move that leads to the subgame G_l where $v_l = \max_{k \in K} v_k$ and continues with the pure strategy s_l^1 in G_l. Similarly, if player 2 starts the game, the value is $\min_{k \in K} v_k$, and the construction is dual to the previous case.   □

6.2.6 Subgame-Perfect Equilibrium

The previous results and proofs extend to multiplayer games. If σ is a strategy in a perfect information game, then for each position p, σ naturally induces a strategy in the subgame $G[p]$ that starts at p (because at every node that follows p, σ prescribes a move).

Definition 6.2.6 A strategy profile σ is *subgame-perfect* if for each position p, the continuation strategy $\sigma[p]$ induced by σ is a Nash equilibrium of $G[p]$.

Theorem 6.2.7 ([187]) *Every finite perfect information game (with or without Nature) has a subgame-perfect equilibrium in pure strategies.*

Proof By backward induction, as in the previous proof.   □

Let us solve by backward induction the following perfect information game (Fig. 6.5):

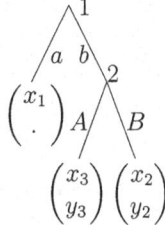

Fig. 6.5 Backward induction in a perfect information game Γ_2

The reduced normal form of Γ_2 is:

	A	B
a	$x_1, \cdot$	$x_1, \cdot$
b	x_3, y_3	x_2, y_2

At the decision node of player 2, he chooses A if $y_3 > y_2$. Suppose this is the case. Player 1 chooses b if $x_3 > x_1$ and (b, A) is the unique subgame-perfect equilibrium. Note, however, that as soon as $x_1 > x_2$ the pair (a, B) is a Nash equilibrium (but is not subgame-perfect). This equilibrium is not self-enforcing: if player 2 has to play, his rational choice is A. Strategy B is called a non-credible threat.

Proposition 6.2.8 *A finite game with perfect information (with or without Nature) with K terminal nodes and I players has generically (with respect to the Lebesgue measure on $\mathbb{R}^{NK}$) a unique subgame-perfect equilibrium.*

Proof If the game does not contain Nature and all payoffs are pairwise distinct, a player is never indifferent during the backward induction process. When Nature is active, by eventually perturbing payoffs at terminal nodes, no player will be indifferent in the process.   □

Remark 6.2.9 The last proposition does not hold for Nash equilibria: in general, there may not be finitely many. However, generically there are finitely many Nash equilibrium payoffs, but not necessarily an odd number [111]. The following example has infinitely many Nash equilibria, two Nash connected components, two equilibrium payoffs, and all equilibria are stable when perturbing payoffs of the extensive form (Fig. 6.6).

Here (b, α) is the unique subgame-perfect equilibrium and $\{(a, x\alpha + (1 - x)\beta) : x \in [\frac{1}{2}, 1]\}$ is a connected component of equilibria where player 2 is playing with a high probability the non-credible threat β.

Remark 6.2.10 The next example (Fig. 6.7) shows that there is no link between backward induction and Pareto-optimality. The unique subgame-perfect equilibrium is $[(L, \ell); T]$ inducing a payoff of $(1, 1)$ while $[(R, r); B]$ is a Nash equilibrium with payoff $(2, 2)$.

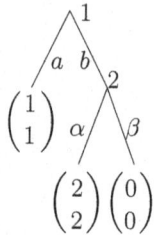
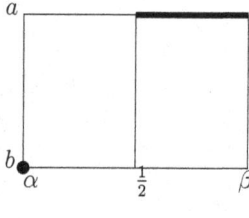

Fig. 6.6 Generic equilibria in extensive form games

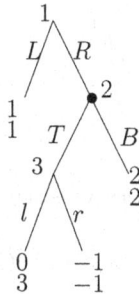

Fig. 6.7 Subgame-perfection and Pareto optimality

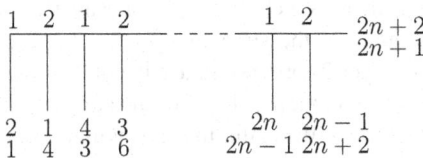

Fig. 6.8 The centipede game

The game of Fig. 6.8 is the famous centipede [181]. Playing horizontal means continue the game, playing vertical means stop. The unique backward induction equilibrium is to stop at every decision node. Thus, at this equilibrium, the game stops immediately, inducing low payoffs (2 for player 1 and 1 for player 2). This leads to a paradox and is in contradiction with laboratory experiments because the payoffs after some amount of cooperation are much larger than immediate defection: if the interaction continues for at least t periods, then both players earn at least t. But payoffs are such that if player i aims to stop at stage t, player $-i$ prefers to stop at stage $t - 1$. This has prompted debate over the concepts involved in the backward induction solutions, see Aumann [6, 7].

6.2.7 Infinite Perfect Information Games

The finiteness in Zermelo's theorem is an essential assumption. For example, the following game in Fig. 6.9 with one player and infinitely many actions does not have an equilibrium (more precisely, does not admit an optimal strategy). However a value exists ($= 1$) and the player has an ε-optimal strategy for every $\varepsilon > 0$ (any action "n" such that $\frac{1}{n} \leq \varepsilon$).

Gale and Stewart [75] introduced the following problem. Two players alternately choose a digit in $\{0, 1\}$. This generates an infinite sequence $x_1, x_2, \ldots$ of zeros and ones that may be interpreted as the binary expansion of a real $x = \sum_{k=1}^{\infty} \frac{x_k}{2^k} \in [0, 1]$. Given a fixed subset $A \subset [0, 1]$, in game G_A player 1 wins if $x \in A$, otherwise player 2 wins.

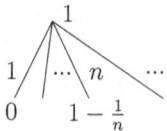

Fig. 6.9 A 1-player perfect information game without 0-equilibrium

Theorem 6.2.11 ([75]) *If A is open or closed, G_A is determined.*

Proof Suppose that A is open and that player 1 does not have a winning strategy. Then, for each choice x_1 of player 1, there is a choice x_2 of player 2 such that player 1 does not have a winning strategy in the subgame $G_A(x_1, x_2)$. Inductively, for every choice x_{2n+1} of player 1, there is a choice x_{2n+2} of player 2 such that player 1 does not have a winning strategy in the subgame $G_A(x_1, x_2, \ldots, x_{2n+2})$. This process defines a strategy τ for player 2 (observe that τ is not defined at nodes excluded by player 2's actions, but this is not a problem because any completion of the strategy will be payoff-equivalent to it). We claim that τ is winning for player 2. Otherwise, there is a winning strategy σ for player 1 against τ. The strategy profile (σ, τ) generates a sequence $x = (x_1, x_2, \ldots) \in A$. Since A is open, there is an n such that $(x_1, x_2, \ldots, x_{2n}, y) \in A$ for every infinite sequence y. Thus, player 1 is winning in the subgame $G_A(x_1, x_2, \ldots, x_{2n})$, a contradiction with the construction of τ.

Suppose A is closed, and that player 2 does not have a winning strategy. Then, there is a choice x_1 of player 1 such that player 2 does not have a winning strategy in the subgame $G_A(x_1)$. But this subgame is open for him. This subgame is determined by the first part of the proof.   □

Theorem 6.2.12 ([75]) *There exists a set A such that G_A is not determined.*

Proof Fix a strategy of player 1. Then the set of $x \in [0, 1]$ that can be generated by strategies of player 2 constitutes a perfect set (i.e. non-empty, closed and dense in itself). A theorem of Bernstein (implied by the axiom of choice) proves that there is a partition of $[0, 1]$ into two sets such that none of them includes a perfect set. It suffices to pick for A any of the elements of a Bernstein partition of $[0, 1]$. If player 1 (resp. player 2) has a winning strategy, then A (resp. the complement of A) includes a perfect set, a contradiction.   □

Martin [126] proved that the perfect information game G_A is determined for every Borel set A. This has deep consequences in descriptive set theory, for example that Borel sets in Polish spaces have the perfect set property. It is also important in logic. The existence of a winning strategy for player 1 reads:

$$(Q_1) \qquad \exists x_1, \forall x_2, \exists x_3, \forall x_4, \ldots, \quad (x_1, x_2, \ldots) \in A$$

If the game is determined then, $\text{Not}(Q_1) = Q2$ where:

$$(Q_2) \qquad \forall x_1, \exists x_2, \forall x_3, \exists x_4 \ldots, \quad (x_1, x_2, \ldots) \notin A.$$

6.3   Extensive Form Games with Imperfect Information

In poker, a player does not know the hands of their adversaries, and in independent move games, as studied in the previous chapters, when a player chooses his strategy he does not know what the other players will do. To include such games in the extensive form model, one needs to add a new object to the description of the game.

6.3.1   Information Sets

The information of player i is represented by a partition $\{P_k^i\}_{k \in K^i}$ of his set of decision nodes $P^i = \bigcup_{k \in K^i} P_k^i$. Each element P_k^i of the partition is called an information set. Nodes in the same information set cannot be distinguished by the player, and so, have the same number of successors, which should correspond to the same sets of physical actions. Each equivalence class of successors is called an *action* and we denote by $A^i(P_k^i)$ the set of actions available to player i at information set P_k^i.

   The first thing to note is that there are different extensive forms to represent a given normal form. For example, the Matching Pennies game has, at least, two representations (depending on which player starts) (Fig. 6.10).

   The fact that player 1 starts the game is a pure convention. Player 2 does not know the choice of player 1 when he plays. In some extensive form games, it is even not possible to order information sets using a public clock. In the game of Fig. 6.11, the history is: player 1 has the choice between calling player 2 or player 3 for an invitation. If $i \in \{2, 3\}$ is called and he/she refuses, then player 1 next calls $j = 5 - i$ without revealing to him/her that i was already called. Thus, when player 2 (resp.

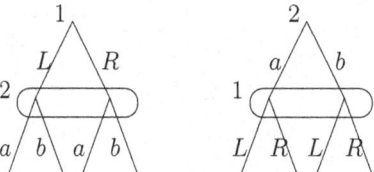

Fig. 6.10 Two extensive form

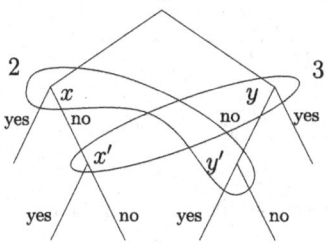

Fig. 6.11 No public clock to order information sets

player 3) receives a call, he/she does not know if he/she was the first or the second choice.

In Fig. 6.11 position x' is in the timing after x and, at the same time y' is after y. But because y and x' belong to the same information set, they should be reached at the same public time and similarly for (x, y').

6.3.2 The Normal Form Reduction

A pure strategy for player $i \in I$ is a mapping that associates to each information set P_k^i of player i *an action* (which is formally an equivalence class of successors) $a^i \in A^i(P_k^i)$.

A profile of pure strategies induces a unique outcome in R. Hence, to every extensive form game is associated a normal form (also called a strategic form). Note that extensive form games with the same normal form may look different, as the following example shows.

All the extensive form games in Fig. 6.12 have the following normal form:

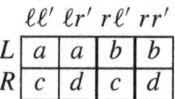

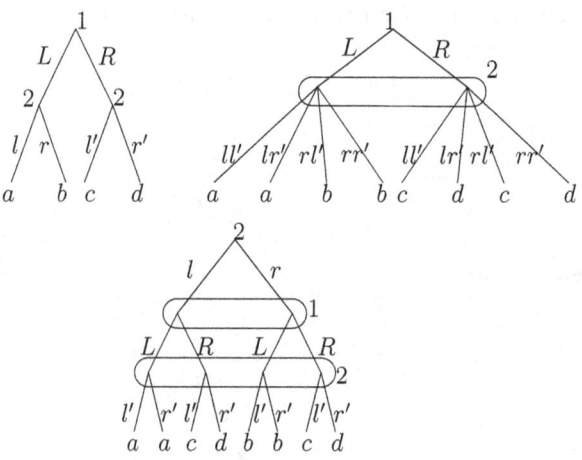

Fig. 6.12 Different extensive forms with the same normal form

More generally, it may be shown [52, 210] that two extensive forms having the same normal form can be linked by a chain of three elementary transformations and their inverses:

(1) interchange of simultaneous moves (as in Matching Pennies above);
(2) coalescing of moves (if a player has two consecutive decision nodes, they can be collapsed to one decision node);
(3) addition of a superfluous move (we add to some decision nodes a redundant action).

This has led some game theorists (starting with Kohlberg and Mertens [108]) to argue that the solution of a game must depend only on the normal form, meaning that two extensive form games with the same normal form must have the same solutions.

6.3.3 Randomized Strategies

Since any simultaneous move game may be represented in extensive form, there is no hope for the existence of a pure Nash equilibrium in all finite extensive form games: randomizations are necessary. However, due to the extensive game structure, there are three different ways for a player to randomize:

(1) A *mixed strategy* is a probability distribution over the set of pure strategies S^i. The interpretation is that, before the game starts, a player selects at random a pure strategy s^i then follows it.
(2) A *behavioral strategy* $\beta^i = \{\beta_k^i\}_{k \in K^i}$ of player i associates to each information set P_k^i a probability distribution $\beta_k^i \in \Delta(A^i(P_k^i))$ on the set of available actions at P_k^i. The interpretation is: a player randomizes step by step, at each time he has to play, among the available actions.
(3) A *general strategy* is a probability distribution over the set of behavioral strategies. The interpretation is, before the game starts, a player selects at random one behavioral strategy β^i and then follows it.

The set of mixed strategies of player i is denoted by $\Sigma^i = \Delta(S^i)$, that of behavioral strategies by $\mathcal{B}^i$ and that of general strategies by $\mathcal{G}^i = \Delta(\mathcal{B}^i)$.

In the following one player problem (Fig. 6.13, the absent-minded driver), the game starts at θ. The decision-maker is amnesic: after choosing b he forgot that he had already played. There are two pure strategies: a or b. They induce the outcomes x and z, respectively. Thus, y is never reached under a pure strategy, hence also under a mixed strategy.

A behavioral strategy in the game Fig. 6.13 is any probability $ta + (1 - t)b$, $t \in [0, 1]$. This induces the probability distribution $tx + t(1 - t)y + (1 - t)^2 z$ on the set of outcomes $R = \{x, y, z\}$. In particular, when $t = \frac{1}{2}$, y is reached with positive probability $\frac{1}{4}$.

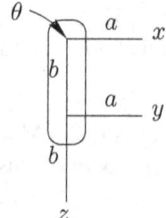

Fig. 6.13 Some behavioral strategies cannot be replicated by any mixed strategy

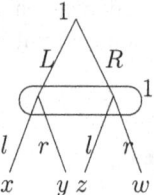

Fig. 6.14 Some mixed strategies cannot be replicated by any behavioral strategy

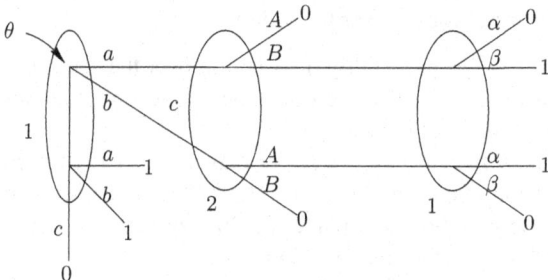

Fig. 6.15 Isbell's game

In the example of Fig. 6.14, again, there is only one player. He controls two information sets, in each of which he has two actions. He thus has four pure strategies: $S = \{Ll, Lr, Rl, Rr\}$. The use of mixed strategies generate all probability distributions over the set of outcomes $R = \{x, y, z, w\}$. On the other hand, a behavioral strategy is defined by the probability $s \in [0, 1]$ to play L and the probability $t \in [0, 1]$ to play l. This induces a probability distribution P over the set of outcomes that satisfies $P(x)P(w) = P(y)P(z) = t(1 - t)s(1 - s)$. Thus, behavioral strategies cannot generate all probability distributions over R.

In the game of Fig. 6.15 [103], pure strategies guarantee 0 to player 1, with behavioral strategies player 1 can guarantee 25/64, with mixed strategies he can guarantee 1/2, and with general strategies he can guarantee 9/16.

6.3.4 Perfect Recall

The above examples show that, in general, the set of mixed and behavioral strategies are not comparable, and both are smaller than the set of general strategies. One introduces here conditions on a game under which they are equivalent.

Definition 6.3.1 An extensive form game is *linear* for player i if there is no play that intersects any of his information sets more than once.

Games in Figs. 6.13 and 6.15 are not linear. The following theorem shows that when a game is linear, any behavioral strategy is outcome equivalent to a general strategy.

Theorem 6.3.2 ([103]) *If an extensive form game is linear for player i then, given any behavioral strategy β^i of player i, there exists a mixed strategy σ^i such that for every general strategy θ^{-i} of players $-i$, the probability distributions $\mathbf{P}(\beta^i, \theta^{-i})$ and $\mathbf{P}(\sigma^i, \theta^{-i})$ on the set of terminal nodes R coincide.*

Proof Every behavioral strategy β^i generates a mixed strategy σ^i as follows. For every pure strategy s^i of player i we let

$$\sigma^i(s^i) = \prod_{k \in K^i} \beta^i[P_k^i, s^i(P_k^i)],$$

where the product is over the family of information sets of player i and $\beta^i[P_k^i, a^i]$ is the probability that, at P_k^i, the action a^i is selected according to the behavioral strategy β^i.

Fixing a pure strategy s^{-i} of players $-i$, the strategy s^i induces an outcome denoted by $r(s^i, s^{-i})$ and the associated play intersects information sets $(P_k^i)_{k \in \widetilde{K}^i}$ for some $\widetilde{K}^i \subset K^i$. The probability of $r(s^i, s^{-i})$ under β^i is

$$\prod_{k \in \widetilde{K}^i} \beta^i[P_k^i, s^i(P_k^i)]$$

and the probability of $r(s^i, s^{-i})$ under σ^i is

$$\sum_{\{\tilde{s}^i \in S^i : r(\tilde{s}^i, s^{-i}) = r(s^i, s^{-i})\}} \sigma^i(\tilde{s}^i) = \sum_{\{\tilde{s}^i \in S^i : \tilde{s}^i(P_k^i) = s^i(P_k^i), \forall k \in \widetilde{K}^i\}} \sigma^i(\tilde{s}^i)$$

$$= \sum_{\{\tilde{s}^i \in S^i : \tilde{s}^i(P_k^i) = s^i(P_k^i), \forall k \in \widetilde{K}^i\}} \prod_{k \in K^i} \beta^i[P_k^i, \tilde{s}^i(P_k^i)],$$

and since each information set appears at most once in $\widetilde{K}^i$ (the game is linear):

$$= \prod_{k \in \widetilde{K}^i} \beta^i[P_k^i, s^i(P_k^i)] \left\{ \sum_{\{\tilde{s}^i \in S^i : \tilde{s}^i(P_k^i) = s^i(P_k^i), \forall k \in \widetilde{K}_i\}} \prod_{k \notin \widetilde{K}^i} \beta^i[P_k^i, \tilde{s}^i(P_k^i)] \right\}.$$

On the other hand,

$$\sum_{\{\tilde{s}^i \in S^i : \tilde{s}^i(P_k^i)=s^i(P_k^i), \forall k \in \widetilde{K}^i\}} \prod_{k \notin \widetilde{K}^i} \beta^i[P_k^i, \tilde{s}^i(P_k^i)]$$

$$= \sum_{k \notin \widetilde{K}^i} \sum_{a^i \in A^i(P_k^i)} \prod_{k \notin \widetilde{K}^i} \beta^i[P_k^i, a^i].$$

The latter sum is equal to 1 by induction on the cardinality of $\widetilde{K}^i$ and because, $\forall P_k^i$, $\sum_{a^i \in A^i(P_k^i)} \beta^i[P_k^i, a^i] = 1$. Consequently, the probability of $r(s^i, s^{-i})$ under σ^i and β^i is the same. This equality of probabilities that holds for every pure strategy s^{-i} of the opponents extends, by payoff-linearity, to every general strategy of $-i$. $\square$

An extensive form game is of perfect recall for a player if, at each of his information sets, he remembers what he did or knew in the past. Formally:

Definition 6.3.3 An extensive form game is of *perfect recall* for player i if for every pair of nodes x and y in the same information set P_k^i, if x'—an (iterated) predecessor of x—belongs to the information set $P_{k'}^i$ then:

– there is a y'—an (iterated) predecessor of y—such that $y' \in P_{k'}^i$;
– the action(s) that leads from x' to x belongs to the same equivalence class as the one(s) that leads from y' to y.

If a game is of perfect recall, it is linear. The converse does not hold, as Fig. 6.12 shows.

Theorem 6.3.4 ([112]) *If the extensive form game is of perfect recall for player i, then for every mixed strategy σ^i of i, there is a behavioral strategy β^i such that for every general strategy θ^{-i} of the other players, the probability distributions $\mathbf{P}(\beta^i, \theta^{-i})$ and $\mathbf{P}(\sigma^i, \theta^{-i})$ on the set of play (terminal nodes) coincide.*

Proof Let σ^i be a mixed strategy. An information set P_k^i is called reachable under σ^i if there is at least one pure strategy s^{-i} of players $-i$ such that the probability of reaching P_k^i under (σ^i, s^{-i}) is positive. Denote this set by $\mathrm{Rch}^i(\sigma^i)$. The behavioral strategy β^i generated by a mixed strategy σ^i is defined as follows:

$$\beta^i(P_k^i; a^i) = \frac{m^i(P_k^i; a^i)}{m^i(P_k^i)},$$

where

$$m^i(P_k^i) = \sum_{s^i : P_k^i \in \mathrm{Rch}^i(s^i)} \sigma^i(s^i) \text{ and}$$

$$m^i(P_k^i; a^i) = \sum_{s^i : P_k^i \in \mathrm{Rch}^i(s^i), s^i(P_k^i)=a^i} \sigma^i(s^i).$$

As in the previous proof, it is sufficient to prove equality of probabilities on R against all pure strategies of the opponents. So, fix some pure strategy s^{-i} of players $-i$.

Let $\overrightarrow{a} = a_1^{j_1}, \ldots, a_T^{j_T}$ be any sequence of actions where player j_t chooses action $a_t^{j_t}$ at time $t = 1, \ldots, T$. Let $\{Q_{t_l}^i\}_{l=1,\ldots,L}$ be the sequence of information sets of player i that crosses $\overrightarrow{a}$. Perfect recall implies that going from $Q_{t_l}^i$ to $Q_{t_{l+1}}^i$ is completely determined by the action $a_{t_l}^i$ and so $\beta^i(Q_{t_l}^i, a_{t_l}^i) = m^i(Q_{t_{l+1}}^i)$. Consequently, $\prod_{l=1}^L \beta^i(Q_{t_l}^i, a_{t_l}^i)$ is a telescopic product that is equal to $m^i(Q_{t_L}^i, a_{t_T}^i)$: the probability of $\overrightarrow{a}$ under (β^i, s^{-i}) and (σ^i, s^{-i}) is the same. □

6.3.5 Nash Equilibrium in Behavioral Strategies

In general, the dimension of the set of behavioral strategies $\mathcal{B}^i$ is much lower than the dimension of the set of mixed strategies Σ^i. If $(P_k^i)_{k \in K^i}$ is the collection of information sets of player i then the dimension of Σ^i is $(\prod_{k \in K^i} |A^i(P_k^i)|) - 1$ and the dimension of $\mathcal{B}^i$ is $\sum_{k \in K^i} (|A^i(P_k^i)| - 1)$ (where $|X|$ is the cardinality of X). Thus, if player i has n information sets, each with two actions, the dimension of $\mathcal{B}^i$ is n and that of Σ^i is $2^n - 1$. Consequently, it is much more efficient to compute Nash equilibria in behavioral strategies. But then one would like to know if they exist and to have a test certifying that a behavioral strategy profile is an equilibrium.

Let Γ be a finite extensive form game of perfect recall, G be its associated normal form game, $\widetilde{G}$ be the mixed extension of G and let $\widetilde{\Gamma}$ be the normal form compact-continuous game where the strategy set of player $i \in I$ is his set of behavioral strategies $\mathcal{B}^i$ in Γ, and i's payoff function in $\widetilde{\Gamma}$ associated to the profile $\beta = (\beta^i)_{i \in I}$ is his expected payoff in Γ.

Definition 6.3.5 We call a Nash equilibrium of $\widetilde{G}$ a *mixed equilibrium* of Γ and a Nash equilibrium of $\widetilde{\Gamma}$ a *behavioral equilibrium* of Γ.

Theorem 6.3.6 *For every finite extensive form game with perfect recall, any mixed equilibrium is outcome equivalent to a behavioral equilibrium and vice-versa, in particular behavioral equilibria always exist.*

Proof Let $m = (m^i)_{i \in I}$ be a mixed equilibrium, which exists by Nash's Theorem since the game is finite. By Kuhn's Theorem 6.3.4, each m^i has an outcome equivalent behavioral strategy β^i. Then we claim that $\beta = (\beta^i)_{i \in I}$ is a Nash equilibrium of $\widetilde{\Gamma}$ (the game in behavioral strategies). Otherwise, there is a player and a behavioral strategy $\widetilde{\beta}^i$ of player i such that $g^i(\widetilde{\beta}^i, \beta^{-i}) > g^i(\beta^i, \beta^{-i})$. But the game is of perfect recall, hence linear, and so by Isbell's Theorem 6.3.2, there is a mixed strategy $\widetilde{m}^i$ of player i outcome equivalent to $\widetilde{\beta}^i$. Consequently, $g^i(\widetilde{m}^i, m^{-i}) > g^i(m^i, m^{-i})$, a contradiction. This shows that any Nash equilibrium of $\widetilde{G}$ is outcome equivalent to a Nash equilibrium of $\widetilde{\Gamma}$. A similar argument shows the converse. □

Definition 6.3.7 An information set Q is *reached* by a behavioral strategy β (or is in the path of β), denoted $Q \in \text{Rch}(\beta)$, if Q has positive probability under β.

Observe that when $Q \in \text{Rch}(\beta)$, using *Bayes' rule*, one can uniquely define a *conditional probability* over nodes of Q, denoted $v_\beta(Q)$.

Theorem 6.3.8 ([213]) *A profile β is a behavioral equilibrium of an extensive form game Γ of perfect recall if and only if for any player i and any information set $Q^i \in \text{Rch}(\beta)$ of player i, $\beta^i(Q^i)$ is a best response of player i in the "local game" starting at Q^i, where Nature chooses a node p in Q^i according to $\nu_\beta(Q)$ and players play in the continuation following what the behavioral profile β prescribes.*

The idea of the proof is simple: if a behavioral profile β is not an equilibrium, then at some reached information set, a player has a profitable deviation (since changing a strategy off-path does not affect payoffs). Conversely, if at some reached information set there is a profitable deviation, then the behavioral profile clearly cannot be a Nash equilibrium.

6.4   Equilibrium Refinement in an Extensive Form Game

Theorem 6.3.8 says that a behavioral strategy profile is a Nash equilibrium if and only if it prescribes best replies along paths consistent with it (but no condition is imposed off path). This is the cause of existence of unreasonable equilibria because one may ask for some properties of the behavior of the players off path. For example, in the game of Fig. 6.6, player 1 playing a and player 2 playing β is a Nash equilibrium while β would not be rational if that part of the game could be reached. Refinement in extensive form games impose rationality on and off path and thus asks for more than playing a best response.

6.4.1   Subgame-Perfect Equilibrium

In some extensive-form games, one can identify parts of the game that have all the properties of an extensive-form game. They are called subgames. More formally, let Γ be an extensive form game with perfect recall. Call a node y a follower of x if there is a sequence of actions from x to y.

Definition 6.4.1 A node x is called a *subroot* if (1) it is the unique element of the information set containing it, and (2) for all information sets Q, all nodes $y \in Q$ are followers of x or no such y is a follower. If x is a subroot, its followers together with the associated information sets and payoffs define a subgame.

Thus, once a subroot is reached, all players know the history that leads to it and, once the game is of perfect recall, at any moment of the game, all players remember this common history. A subgame of an extensive form game Γ is proper if it is different from Γ. For example, in the game described in Fig. 6.16, there are two proper subgames.

Definition 6.4.2 A *subgame-perfect equilibrium* (SPE) is a behavioral strategy profile that prescribes a Nash equilibrium in every subgame.

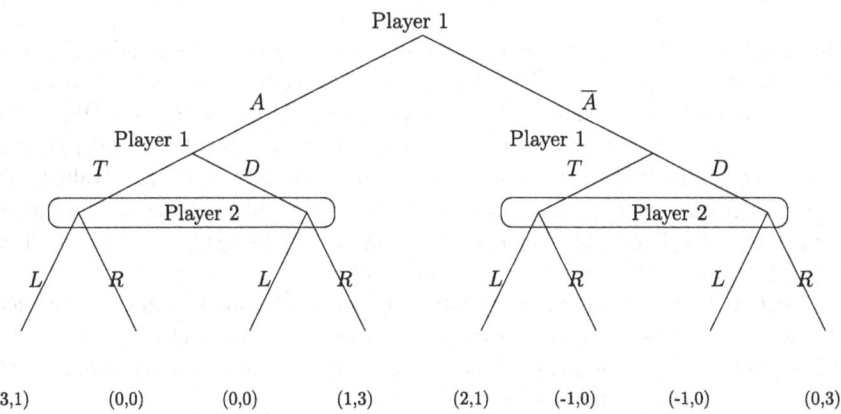

Fig. 6.16 The extensive form of the burning money game

Theorem 6.4.3 ([187]) *Every extensive form finite game with perfect recall admits a subgame-perfect equilibrium.*

Proof The proof is by induction on the number of subgames. If the game has no proper subgame then Nash equilibria in behavioral strategies exist by Theorem 6.3.6. Otherwise, all proper subgames $\{\Gamma_{x_1}, \ldots, \Gamma_{x_k}\}$ are of perfect recall and have a smaller number of subgames and so, by assumption, admit subgame-perfect equilibria $\{\beta_1, \ldots, \beta_k\}$, with associated payoffs $\{g_1, \ldots, g_k\}$. Create a new game played as in Γ, and terminate at nodes $\{x_1, \ldots, x_k\}$ with terminal payoffs $\{g_1, \ldots, g_k\}$. This game has no proper subgame and so, by Theorem 6.3.6, has a behavioral equilibrium β_0. Concatenating $\beta_0, \beta_1, \ldots, \beta_k$ yields a subgame-perfect equilibrium for Γ. □

The game of Fig. 6.16 is an equivalent representation of the "Burning Money" game introduced by Ben-Porath and Dekel [21]. Player 1 can play a "battle of the sexes" by choosing A or can first sacrifice one unit of payoff and then play the same "battle of the sexes" (playing $\bar{A}$).

A translation of player 1's payoff by one does not change the set of Nash equilibria and so, both subgames have the same set of Nash equilibria: two pure (T, L), (D, R) and one mixed $(\frac{3}{4}T + \frac{1}{4}D; \frac{1}{4}L + \frac{3}{4}R)$. In subgame A, equilibrium payoffs are $(3, 1)$, $(1, 3)$ and $(\frac{3}{4}, \frac{3}{4})$ respectively and in subgame $\bar{A}$ payoffs are decreased by 1 for player 1 and so are $(2, 1)$, $(0, 3)$ and $(-\frac{1}{4}, \frac{3}{4})$. Any concatenation of an equilibrium in subgame A and an equilibrium in game $\bar{A}$ generates exactly one, or infinitely many, subgame-perfect equilibria in the entire game (depending on whether player 1 is indifferent or not). Since player 1's equilibrium payoffs in different subgames are pairwise distinct, "backward induction" generates only finitely many equilibria in this example. For example, if in subgame A players follow (D, R) and in subgame $\bar{A}$ they follow (T, L), then player 1's unique optimal choice is $\bar{A}$ (he obtains 2 instead of 1). Computing subgame equilibrium payoffs is quicker.

6.4.2   Sequential and Bayesian Perfect Equilibria

The next two refinements refine subgame-perfection by imposing a rational behavior on all information sets and not only on subroot nodes. To do so, one needs to associate to each information set Q^i a probability distribution $\mu(Q^i)$ over the nodes $\{p_1, \ldots, p_L\}$ of Q^i. The collection of all such probabilities is called a *belief system*. Given $\mu(Q^i)$ and the continuation strategy induced by β, player i can evaluate his payoff for any action $a^i \in A^i(Q^i)$, and sequential rationality imposes that his payoff is maximized by $\beta^i(Q^i)$. Of course, the belief system $Q \to \mu(Q)$ must be consistent with the strategy profile β, as the following explains.

Recall that an information set Q is *reached* by a behavioral strategy β, denoted $Q \in \text{Rch}(\beta)$, if Q has positive probability under β and when $Q \in \text{Rch}(\beta)$, using *Bayes' rule*, one can uniquely define a *conditional probability* over nodes of Q, denoted $v_\beta(Q)$ (See Definition 6.3.7).

Definition 6.4.4 ([71]) A pair (β, μ) consisting of a behavioral strategy profile β and a belief system μ is weak Bayesian perfect if:

- for any information set Q^i, $\beta^i(Q^i)$ is a best response of player i in the game starting at Q^i where Nature chooses a node p in Q^i according to $\mu(Q^i)$ and players play in the continuation game following the strategy profile β.
- μ is Bayes-compatible with β in the sense that for every $Q \in \text{Rch}(\beta)$, $\mu(Q) = v_\beta(Q)$.

This concept does not impose any belief restriction on information sets that are not reached by β. The next refinement adds restrictions. Let $\mathcal{B}^i$ be the set of behavioral strategies of player i and $\mathcal{B} = \prod_i \mathcal{B}^i$. A behavioral strategy profile is interior (or completely mixed) if all actions are played with positive probability at every information set. Denote the corresponding sets by $\text{int}\mathcal{B}^i$ and $\text{int}\mathcal{B} = \prod_i \text{int}\mathcal{B}^i$.

Let Φ be the correspondence that associates to every behavioral profile $\beta \in \mathcal{B}$ the family of belief systems that are Bayes-compatible with β. Thus, when $\beta \in \text{int}\mathcal{B}$, Φ is single-valued: $\Phi(\beta) = \{v_\beta\}$ (because all information sets are reached with positive probability). The example of Fig. 6.17 shows that

$$\overline{\Phi(\text{int}\mathcal{B})} \neq \Phi(\mathcal{B}).$$

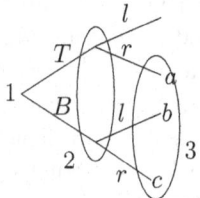

Fig. 6.17 Beliefs resulting from interior strategies form a strict subset of all beliefs

Actually, if player 1 plays $(1 - \varepsilon)T + \varepsilon B$ and player 2 plays $(1 - \varepsilon)l + \varepsilon r$, the total probability over the set of nodes $\{a, b, c\}$ that constitutes the unique information set of player 3 is $(1 - \varepsilon)\varepsilon a + \varepsilon(1 - \varepsilon)b + \varepsilon^2 c$, so that the conditional probability of node c being reached, $\frac{\varepsilon^2}{2(1-\varepsilon)\varepsilon+\varepsilon^2}$, goes to zero as ε goes to zero. Hence $(1/2, 1/2, 0)$ belongs to the closure $\overline{\Phi(\text{int}\mathcal{B})}$. However, if a and b have positive probabilities, then T and B have positive probabilities, as do ℓ and r, and hence also c, thus $(1/2, 1/2, 0) \notin \Phi(\mathcal{B})$.

Definition 6.4.5 ([111]) A pair (β, μ), $\beta \in \mathcal{B}$ and $\mu \in \Phi(\beta)$ is a *sequential equilibrium* if it is weak Bayesian perfect and there exists a sequence $\beta_n \in \text{int}\mathcal{B}$ converging to β such that $v_{\beta_n} \to \mu$.

The justification is that players can tremble and all mistakes have positive probability (β_n is interior) but the probability of a mistake is very small (β_n is close to β).

Actually, backward induction computations, if justified by common knowledge of rationality, rapidly yields to contradictions: if some decision nodes are not supposed to be reached, then why should a player continue to believe in the rationality of the opponents [173]? In the centipede game described above (Fig. 6.8), common knowledge of rationality implies that at any stage, every player anticipates that his opponent will stop at the next stage, yielding him to stop. But what should player 2 believe at stage 2 if he observes that player 1 continues at stage 1? Why should he believe that player 1 will behave rationally later? One way to fix the problem is by assuming that players can make mistakes (Selten's trembling hand). Hence, if the play reaches a node which is not supposed to be reached, this is only by mistake.

The extensive form game in Fig. 6.17 shows that sequential equilibrium is a strict refinement of weak Bayesian perfect equilibrium.

Theorem 6.4.6 ([111]) *The set of sequential equilibria of a finite extensive form game with perfect recall is non-empty.*

Proof See Sect. 6.5 below.   □

6.5   Equilibrium Refinement in Normal Form Games

Let $G = (I, \{S^i\}_{i \in I}, \{g^i\}_{i \in N})$ be a finite normal form game, denote by $\Sigma^i = \Delta(S^i)$ the set of mixed strategies of player i, let $\Sigma = \prod_i \Sigma^i$ be the set of mixed strategy profiles, and let $\text{int}\Sigma$ be the set of completely mixed strategies σ (i.e. $\text{supp}(\sigma^i) = S^i$, $\forall i \in I$).

Definition 6.5.1 For any $\varepsilon > 0$, a completely mixed strategy profile $\sigma_\varepsilon \in \text{int}\Sigma$ is ε-*perfect* if for every player $i \in I$ and every pure strategy $s^i \in S^i$, if s^i is not a best response against σ_ε^{-i}, then $\sigma_\varepsilon^i(s^i) \leq \varepsilon$.

The interpretation is: all strategies can potentially be played, but strategies that are sub-optimal are played with a very small probability.

Definition 6.5.2 A strategy profile $\sigma \in \Sigma$ is a *perfect equilibrium* if there is a sequence $\sigma_n \in \operatorname{int} \Sigma$ of ε_n-perfect equilibria that converges to σ as ε_n goes to 0.

Theorem 6.5.3 ([187]) *Every finite normal form game has a perfect equilibrium, and every perfect equilibrium is a Nash equilibrium.*

Proof For every $\tau \in \operatorname{int} \Sigma$ and every $\varepsilon \in]0, 1[$, define the *perturbed game* $G(\tau; \varepsilon)$ played as in G but with the modified payoff $\sigma \mapsto g((1 - \varepsilon)\sigma + \varepsilon\tau)$. By Nash's theorem, this game has a Nash equilibrium v_ε, and $\sigma_\varepsilon = (1 - \varepsilon)v_\varepsilon + \varepsilon\tau$ is ε-perfect. By compactness of the strategy space, every sequence of ε_n-perfect equilibria has a converging subsequence. Finally, by continuity of the payoffs, non-best response strategies have a zero probability under perfect equilibrium, and a perfect equilibrium is a Nash equilibrium.   □

This notion eliminates *dominated equilibria* (i.e. equilibria that contains in their support a weakly dominated strategy). Actually, such a strategy is sub-optimal for any ε-perfect equilibrium and so is played with probability at most ε. Thus, its probability vanishes as ε goes to zero.

Proposition 6.5.4 ([215]) *In a finite two-player normal form game, an equilibrium is perfect if and only if it is undominated. The inclusion is strict for more-than-two-player games.*

In the game of Fig. 6.6, the unique perfect equilibrium of the normal form is the subgame-perfect equilibrium of the extensive form. Thus, in this example, normal form perfection allows us to detect the backward induction equilibrium. This is not true in every extensive form game. The following refinement can do.

Definition 6.5.5 For any $\varepsilon > 0$, a completely mixed strategy profile $\sigma_\varepsilon \in \operatorname{int} \Sigma$ is ε-*proper* if for every player $i \in I$ and every two pure strategies s^i and t^i in S^i, if $g^i(s^i, \sigma_\varepsilon^{-i}) < g^i(t^i, \sigma_\varepsilon^{-i})$, then $\sigma_\varepsilon^i(s^i) \le \varepsilon \, \sigma_\varepsilon^i(t^i)$.

In this notion, players take rationally into account the fact that they can make mistakes and so more costly mistakes are played with an infinitely smaller probability. Observe that when a strategy is sub-optimal then it is played with a probability of at most ε. Consequently, an ε-proper equilibrium is also ε-perfect.

Definition 6.5.6 A strategy profile $\sigma \in \Sigma$ is a *proper equilibrium* if it is the limit of a sequence $\sigma_n \in \operatorname{int} \Sigma$ of ε_n-proper equilibria as ε_n goes to 0.

Theorem 6.5.7 ([148]) *Every finite normal form game has a proper equilibrium and every proper equilibrium is a perfect equilibrium and so is a Nash equilibrium.*

Proof See Exercise 3.   □

Remark 6.5.8 Any strict equilibrium $s \in S$ is proper. This is because, for every small perturbation of the opponents, s^i remains the unique best response of player i. Now, using Myerson's result we know that the restricted game, where each player i cannot use the pure strategy s^i, has a ε-proper equilibrium τ_ε. Consequently, $\sigma_\varepsilon = (1 - \varepsilon)s + \varepsilon\tau_\varepsilon$ is ε-proper and converges to s.

Proper equilibrium is a strict refinement of perfect equilibrium, as the following example shows.

	l	m	r
T	$(1, 1)$	$(0, 0)$	$(-1, -2)$
M	$(0, 0)$	$(0, 0)$	$(0, -2)$
B	$(-2, -1)$	$(-2, 0)$	$(-2, -2)$

M is the unique best response to B and so is not weakly dominated, and similarly for m. Thus (M, m) is undominated and so is perfect (by Proposition 6.5.4). But (M, m) is not proper. Actually, B and r are strictly dominated by T and l respectively, and so, every ε-proper equilibrium that converges to (M, m), T (resp. l) will be infinitely more probable than B (resp. r). Player 1, facing any perturbed strategy of the form $\varepsilon_1 l + (1 - \varepsilon_1 - \varepsilon_1\varepsilon_2)m + \varepsilon_1\varepsilon_2 r$, where $\varepsilon_i \leq \varepsilon$, has as unique best response T and not M. Thus, (M, m) cannot be the limit of ε-proper equilibria.

This example shows that the set of perfect equilibria depends on the addition or deletion of a strictly dominated strategy: if B and d (strictly dominated) are eliminated, (M, m) is no longer a perfect equilibrium. Proper equilibrium has the same problem, as the next example shows (see Fig. 6.20 and the corresponding section for an in depth discussion about this famous "Battle of the Sexes" game with outside option).

	l	r
T	$(2, 4)$	$(2, 4)$
M	$(3, 1)$	$(0, 0)$
B	$(0, 0)$	$(1, 3)$

The strategy profile (T, r) is a proper equilibrium. Actually, T is the unique best response to r and so remains a best response for any small perturbation of r. If T is perturbed as follows: $(1 - \varepsilon - \varepsilon^2)T + \varepsilon^2 M + \varepsilon B$, the unique best response of player 2 is r. Thus, (T, r) is proper. However, the elimination of the strictly dominated strategy B implies that r becomes weakly dominated, and so (T, r) is not perfect, and so not proper.

6.6   Linking Extensive and Normal Form Refinements

Definition 6.6.1 The *agent normal form* Γ^a of an extensive form game with perfect recall Γ is an extensive form game played as in Γ where each information set Q in Γ corresponds to a player $i(Q)$ in Γ^a whose payoff is identical to the payoff of the player who plays at Q in Γ.

Thus, the agent normal form operation consists in duplicating each player, by a team of agents, as many times as he has information sets: all agents of a player have a common interest with him. In this game, each agent plays at most once, and so, formally, this game may be identified with its normal form: mixed and behavioral strategy sets coincide.

Theorem 6.6.2 ([111]) *The set of sequential equilibria of a finite extensive form game with perfect recall Γ is non-empty.*

Proof Consider a Selten perturbed game Γ_ε^a associated to the agent normal form game Γ^a where each agent is restricted to play any of his actions (=strategies) with probability at most ε. This game satisfies the hypothesis of Glicksberg's theorem and so admits a Nash equilibrium σ_ε. This induces an $O(\varepsilon)$-equilibrium in Γ^a where each information set Q is reached with positive probability and where each player $i(Q)$ is optimizing his payoff up to $O(\varepsilon)$ given his belief $\mu_\varepsilon(Q)$ induced by σ_ε. Taking a converging subsequence, one obtains for each information set Q a strategy profile $\sigma(Q)$ together with a belief $\mu(Q)$ where at each information set, agent $i(Q)$ is maximizing his payoff given his belief. This constitutes a sequential equilibrium in Γ.   □

Definition 6.6.3 ([187]) An *extensive form perfect equilibrium* of Γ is a perfect equilibrium of the agent normal form Γ^a.

The previous proof constructs a sequential equilibrium of Γ by taking the limit of extensive form ε-perfect equilibria. One may ask if there is a connection between extensive and normal form perfect equilibria. The following two examples show that they may differ. In the game of Fig. 6.18, $(R\ell; a)$ is extensive form perfect but is not perfect in the associated normal form.

In the game of Fig. 6.19, Bt is perfect in the normal form but is not perfect in the extensive form.

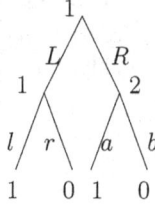

Fig. 6.18 Perfect in the extensive form and not in the normal form

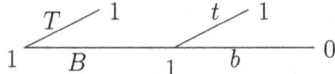

Fig. 6.19 Perfect in the normal form but not in the extensive form

In fact, there are games where the intersection of the two sets is empty. The following result proves, however, that there is a deep link between normal form and extensive form refinements.

Theorem 6.6.4 ([108, 213]) *A proper equilibrium σ of a finite normal form game G induces a sequential equilibrium (β, μ) in every extensive form game with perfect recall Γ having G as normal form.*

Proof By definition, σ is the limit of a sequence of ε_n-proper equilibria $\sigma_n \in \text{int} \Sigma$. From Kuhn's theorem's proof above, the completely mixed strategy profile σ_n generates an equivalent behavioral strategy profile $\beta_n \in \text{int} \mathcal{B}$ which is interior. Thus, every information set Q is reached with positive probability under β_n, inducing a well defined Bayes-compatible belief system $\{\mu_n(Q)\}$. Take a subsequence such that (β_n, μ_n) converges to some (β, μ) (using compactness). Consider some player i and let Q^i be one of his information sets. Let us prove that at Q^i, i is playing a best response given his belief $\mu(Q^i)$ and β. By contradiction, if not, there is some action b^i in $A^i(Q^i)$ not played in $\beta^i(Q^i)$ that performs better than some actions a^i in the support of $\beta^i(Q^i)$. By continuity, the payoff of b^i is strictly better than the payoff of a^i facing β_n when n is large enough. Since σ_n is ε_n-proper, the probability of any pure strategy s^i that plays a^i at Q^i is at most ε_n, the probability of any pure strategy t^i that differs from s^i only at Q^i and plays b^i. Consequently, by construction of β_n, $\beta_n[Q^i; a^i] \leq \varepsilon_n \beta_n^i[Q^i, a^i]$, which implies that $\beta^i[Q^i, a^i] = 0$, a contradiction.   □

This result is remarkable. It shows that backward induction, a property specific to an extensive form description of the game, is detectable in the normal form where the notions of information and timing do not exist.

6.7 Forward Induction and Strategic Stability

In the definition of backward induction and sequential rationality, a player is always looking at the future and his behavior is, at every decision node, independent of the past. An action is rational at some information set if it is a best response given the belief at this position and the anticipated future play. There is no restriction on past play, except that the belief at an information set Q must be strategy-consistent (justified by some sequence of small mistakes). Forward induction, introduced by Kohlberg and Mertens [108], uses past play to restrict the set of possible beliefs off the equilibrium path: whenever possible, a player justifies deviations as a pure rationality act instead of just a mistake. The following famous example in Fig. 6.20 illustrates the idea.

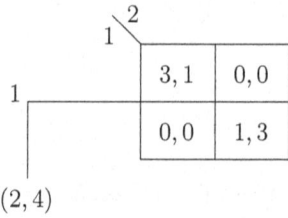

Fig. 6.20 "Battle of the sexes" with an outside option

At the first stage, player 1 has the option to stop the game (by playing down), receiving a payoff of 2, or to continue. If he continues, he plays a battle of the sexes with player 2. The normal form (discussed above in Sect. 6.3.2) is:

	L	R
S	(2, 4)	(2, 4)
T	(3, 1)	(0, 0)
B	(0, 0)	(1, 3)

As shown above, the strategy profile $\sigma = (S, R)$ is a proper equilibrium and so induces a sequential equilibrium in the extensive form described in Fig. 6.19. In the corresponding sequential equilibrium, player 1 does not enter the game and if he enters, player 2 plays R because he believes that player 1 will play B with a high probability. But should player 2 have such a belief? Since player 1 could have obtained 2 by stopping (playing S at the first stage), if he is rational and continues, this means that he anticipates more than 2. But, playing B yields him at most 1 (B is strictly dominated by S). So, the unique belief consistent with player 1's rationality at stage 2 is probability one on T. But then, player 2 should play L, and this induces player 1 to play T. The unique Nash equilibrium compatible with forward induction reasoning is thus (T, L) with payoff $(3, 1)$.

Observe, however, that one may have incompatibility between forward and backward induction principles, as the example in Fig. 6.21 shows. Just add a stage before the previous game.

In the subgame after the first move, the preceding argument shows that the unique forward induction equilibrium is (T, L) with payoff $(3, 1)$. By backward inductionplayer 2 should stop at the first stage, and obtains 2. However, if player 2 contin-

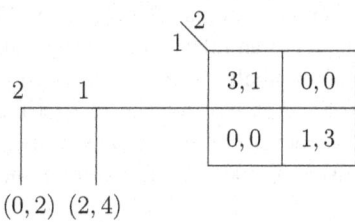

Fig. 6.21 Incompatibility between backward and forward inductions

ues, by the forward induction logic, this means that he expects more than 2, hence in the second subgame he will play R, and so player 1 should play stop in the first subgame, a contradiction. This is why in "the definitions" of forward induction, it is required that this reasoning should be applied only "whenever possible".

To fix the problem and unify backward and forward induction ideas, Kohlberg and Mertens [108] defined a set of axioms that a solution concept should satisfy. The most important are: (1) it must only depend on the reduced normal form, (2) it must be a connected set of non-weakly dominated equilibria, (3) it must contain at least a proper equilibrium, and (4) if a weakly dominated strategy is eliminated, the solution does not vanish (meaning that if S is stable for G and G' is obtained from G after a weakly dominated strategy is eliminated, then G' has a stable solution $S' \subset S$). There are also other properties that we will not list.

The last "axiom" implies that if an outcome survives the sequence of elimination of weakly dominated strategies, it must be stable and if it is the unique surviving outcome, it is the unique stable outcome. For example, in the battle of the sexes with outside option, B is strictly dominated (by S), then R is weakly dominated, then S, which leads to (T, L) (the forward induction outcome). In the burning money game (Fig. 6.16), the unique equilibrium which survives to iterative elimination of weakly dominated strategies is: player 1 plays A and they follow with (T, L) in the battle of the sexes with payoff $(3, 1)$. Why this outcome is the unique compatible with forward induction is easy to understand. Observe that the unique equilibrium in which player 1 does a sacrifice (plays $\bar{A}$) gives him a payoff of 2. If he does not play $\bar{A}$ but instead plays A, this means that he expects more than the payoff 2. The unique equilibrium of the subgame compatible with this is (T, L). That's exactly the situation we had in the battle of the sexes with outside option. This example is interesting because it shows that having the possibility to destroy units of payoffs could help a player to obtain a better equilibrium.

The search for a solution leads to the following definition of strategic stability. A subset C of Nash equilibria of an extensive/normal form game Γ is *strategically stable* if it is connected and if for any normal form game G that has the same *reduced normal form* as Γ and for any game G_ε in a neighborhood of G, G_ε has a Nash equilibrium σ_ε, close to C. In G, some mixtures of strategies in Γ may be added as pure strategies or some pure strategies that are payoff-equivalent to some mixtures are dropped: in that sense G and Γ will have the same reduced normal form. But how to define a neighborhood?

If the perturbation is on the set of payoffs, one obtains essential components (i.e. connected components of Nash equilibria with non-zero index), see Govindan and Wilson [85]. However, this solution concept has the disadvantage of violating the axiom of *admissibility* (i.e. some weakly dominated equilibria are not eliminated):

	L	R
T	(2, 2)	(2, 2)
B	(1, 1)	(0, 0)

In this example, (T, R) is weakly dominated but is strict (and so stable) in the close-by game:

	L	R
T	(2, 2)	(2, 2 + ε)
B	(1, 1)	(0, 0)

A different way to define the neighborhood of a game is with respect to strategy perturbations instead of payoff perturbations. One possibility is by restricting every player i to any polytope $\Sigma^i(\varepsilon)$ of the mixed strategy set such that its complement is within ε from the boundary of Σ^i. Having a polytope guarantees that the perturbed game is equivalent to a finite game (where pure strategies are the extreme points of the polytope). Here also we lose admissibility, as the following example shows:

	L	R	Z
T	(2, 2)	(2, 2)	(0, 0)
B	(1, 1)	(0, 0)	(0, 0)

R becomes admissible if it is less perturbed throughout Z than L.

An alternative approach is to generate the neighborhood of a game by defining for all $(\varepsilon > 0, \sigma \in \text{int}\,\Sigma)$ the perturbed game $G(\sigma; \varepsilon)$ in the sense of Selten. Mertens [133, 134] proved that essential components with respect to this topology are connected, BR-invariant, admissible, satisfy backward induction, forward induction and many other properties. Proving that Mertens stable sets are the only ones that satisfy the good properties is still an open problem, and has been solved only for some particular classes of games (generic two-player extensive form games, [86]). To learn more about strategic stability, see the surveys of Hillas and Kohlberg [98] and of van Damme [217] in the *Handbook of Game Theory*.

6.8 Exercises

Exercise 1. Chomp

Let n and m be two positive integers and define a perfect information game as follows. Two players take alternate turns on an $n \times m$ checkerboard defined by positions (x, y) where $x = 1, \ldots, n$ and $y = 1, \ldots, m$. Each player checks an uncovered square (x, y), which implies covering all squares (x', y') with $x' \geq x$ and $y' \geq y$. The player forced to cover the square $(1, 1)$ loses the game.

(1) Show that the player who makes the first move has a winning strategy.
(2) Give explicitly the winning strategy when $n = m$.
(3) Is the game determined when n or m (or both) are infinite? Which player has a winning strategy?

Exercise 2. A poker game

Two players play a zero-sum poker game. They bet 1 euro each before the game starts. Then, Nature uniformly selects a card in $\{H, L\}$, and only player 1 observes it. Player 1 then decides between S and C (Stop or Continue). If he stops he loses his euro. Otherwise, he bets an additional 1 euro (so, in total he contributes 2 euros). If player 1 chooses C, player 2 has to decide between A and F (Abandon or Follow). If he abandons, he loses his euro, otherwise he adds 1 euro to see the card selected by Nature. If the card is H, player 1 wins the 2 euros of player 2, otherwise, player 2 wins the 2 euros of player 1.

(1) Write the game in extensive and normal forms.
(2) Compute the value and the optimal mixed strategies.
(3) Deduce the optimal behavioral strategies.
(4) Conclude.

Exercise 3. Existence of proper equilibria

Let $G = (I, (S^i)_{i \in I}, (g^i)_{i \in I})$ be a finite game in strategic form. Let us prove the existence of proper and perfect equilibria (see Definitions 6.5.1 and 6.5.5).

Let $\varepsilon \in]0, 1[$ be fixed. For each player $i \in I$, define $\eta^i = \varepsilon^{|S^i|}/|S^i|$, $\Sigma^i(\eta^i) = \{\sigma^i \in \Delta(S^i), \sigma^i(s^i) \geq \eta^i \; \forall s^i \in S^i\}$, and let $\Sigma(\eta) = \prod_{i \in I} \Sigma^i(\eta^i)$.

Define a correspondence:

$$F : \Sigma(\eta) \longrightarrow \Sigma(\eta)$$
$$\sigma \longmapsto \prod_{i \in I} F^i(\sigma)$$

where for $i \in I$ and $\sigma \in \Sigma$,

$$F^i(\sigma) = \{\tau^i \in \Sigma^i(\eta^i), \forall s^i, t^i, (g^i(s^i, \sigma^{-i}) < g^i(t^i, \sigma^{-i})) \Longrightarrow \tau^i(s^i) \leq \varepsilon \tau^i(t^i)\}.$$

(1) Show that F has non-empty values.
(2) Applying Kakutani's theorem, conclude that the game has an ε-proper equilibrium.
(3) Compute all Nash, perfect and proper equilibria (in pure and mixed strategies) of the following games:

	L	R
T	$(1, 1)$	$(0, 0)$
B	$(0, 0)$	$(0, 0)$

	L	M	R
T	$(1, 1)$	$(0, 0)$	$(-9, -10)$
M	$(0, 0)$	$(0, 0)$	$(-7, -10)$
B	$(-10, -9)$	$(-10, -7)$	$(-10, -10)$

Exercise 4. Bargaining

Two players bargain to divide an amount of money $M > 0$. If they fail to reach an agreement, both receive zero (the disagreement payoff). Let us compute subgame-perfect equilibria of several extensive form bargaining games, Rubinstein [182].

(1) *The Ultimatum Game*

Player 1 makes a take it or leave it offer to player 2, namely: at stage 1, player 1 proposes a sharing $x \in [0, M]$ (which means player 2 receives x and player 1 $M - x$). At stage 2, player 2 can Accept or Reject the offer. If he accepts, $(M - x, x)$ is implemented, otherwise, both players receive zero.

Write the game in extensive form and show it has a unique subgame-perfect equilibrium.

(2) *The finitely repeated alternative offer game*

Now the bargaining process can take $2 \leq T < +\infty$ periods and the amount M shrinks from one period to the next by a factor $\delta \in]0, 1[$. Alternately, players make offers and the game goes to the next round if the offer of the actual round is rejected, until stage T when the game terminates. More precisely, inductively, at odd (resp. even) stages player 1 (resp. player 2) makes an offer $x_t \in [0, M_t]$, where $M_t = \delta^{t-1} M$. When player $i = 1, 2$ makes his offer, this means he proposes x_t to player $j = -i$ (and the remaining for him). Player j has the option to accept i's offer or to reject it. If j accepts, the offer is implemented. If $t < T$, the game moves to round $t + 1$. Otherwise, the game is over and both players receive zero.

Find, by induction on T, the subgame-perfect equilibrium payoff and its limit as $T \to \infty$.

(3) *The infinitely discounted repeated alternative offer game*

Consider the previous game with $T = \infty$. Show that the following strategy is the unique subgame-perfect equilibrium. At every stage t, if the game hasn't stopped, the player who plays offers a fraction $\frac{\delta}{1+\delta}$ of the amount M_t at time t to the other player (i.e. $x_t = \frac{\delta}{1+\delta} M_t$) and the other player accepts the offer.

Exercise 5. An extensive form game

Let G be a finite game in extensive form with perfect recall and G' be obtained from G by refining some information sets (dividing them into two or more sets). Let σ be a Nash equilibrium of G.

(1) Show that if σ is pure and there are no chance moves, σ is a Nash equilibrium of G'.

(2) Show that the above result does not extend to σ mixed or with chance moves.

Exercise 6. Stack of tokens a Nim game
There are two players and a stack of n tokens. Players play alternately. At each round, a player can remove 1, 2 or 3 tokens. The player who takes the last token loses. When does the first mover have a winning strategy?

Exercise 7. An entrant with incomplete information
Consider two firms in competition in a market where only one of the two can make a profit. A chance move determines if firm 1 makes a technological innovation I or not N. If it did, firm 1 is able to beat any competitor. This game may be described as follows:

At stage 0:   Nature selects I or P with equal probabilities, only firm 1 is informed.

At stage 1:   Firm 1 decides if it stays in the market (S) or quits it (Q). If it quits, its payoff is 0 and firm 2 enters and makes a benefit of 4.

At stage 2:   Observing that firm 1 stays, firm 2 decides to enter the market (E) and compete with firm 1 or to leave the market (L). If it leaves, firm 1 gets 4 and firm 2 gets 0. Otherwise, the payoff depends on Nature's choice. If firm 1 makes an innovation, it obtains a payoff of 8, and firm 2 a payoff of -4. Otherwise, firm 2 gets 8 and firm 1 makes -4. Observe that this game is constant sum, and so has a value.

(1) Write the game in extensive and normal forms.
(2) Compute the value and the optimal mixed strategies.
(3) Deduce the optimal behavioral strategies.
(4) Conclude.

Exercise 8. Double Auction
A seller (player 1) and a buyer (player 2) trade an indivisible good. The cost for the seller is c and the value for the buyer is v. We suppose that c and v are drawn independently from the uniform distribution on the interval $[0, 1]$. The rule of the auction is as follows. The buyer and the seller simultaneously submit their offers ($b^1 \in [0, 1]$ for player 1 and $b^2 \in [0, 1]$ for player 2). b^1 is interpreted as the minimal price that player 1 will accept in order to sell and b^2 is the maximal price that player 2 will accept in order to buy. If $b^1 > b^2$ there is no trade. Otherwise, they trade at the price $p = \frac{b^1 + b^2}{2}$.

Suppose first that information is complete (i.e. v and c are common knowledge).

(1) Write the normal form and compute all pure Nash equilibria.

Suppose from now on that the information is incomplete: each player knows only his own type (player 1 observes c before playing, and player 2 observes v). Let $\beta_i(\cdot) : [0, 1] \to [0, 1]$, Borel measurable, be a strategy of player i.

(2) Prove that β^i, $i = 1, 2$ are strictly increasing
(3) Assuming that β^i, $i = 1, 2$ are of class C^1, provide the pair of ODEs that they must satisfy if they constitute a Nash equilibrium.
(4) Show that there exists a linear solution to the ODEs of the previous question. Under which condition is there a trade?

6.9 Comments

The literature on extensive form games is extremely large and its impact on applications, in particular in economics, is huge. Research in this field is still in full development.

The link with computer science introduces a lot of questions related to complexity issues:
– What kind of results can be obtained by restricting the players to use subclasses of strategies like having finite memory or being implementable by finite automata?
– What is the complexity of finding a specific solution concept?
– What is the complexity of the solution itself, in terms of information required or of computational cost to implement it?

An important direction of analysis is related to the formalization of the beliefs (hierarchy of knowledge or conditional systems), a field called epistemic game theory [40].

Finally, the research on stability is still very active, see for example [85].

Chapter 7
Correlated Equilibria, Learning, Bayesian Equilibria

7.1 Introduction

Correlated equilibrium is an extension of Nash equilibrium introduced by Aumann [3] in 1974. In this solution, each player may observe, before playing the game, a private signal, generated from some commonly known distribution. This correlation device leads to an increase of the set of equilibria, which becomes a nice efficiently computable polytope, the opposite of the set of Nash equilibria, a semi-algebraic set with several components and whose computational complexity belongs to the PPAD class.

The next section defines correlated equilibrium in its general form and shows the equivalence with its more popular canonical representation.

Section 7.3 introduces the so-called internal and external no-regret criteria and provides two discrete time learning uncoupled procedures that satisfy the no-regret criteria. The proofs are based on Blackwell approachability. Finally, it is shown that any externally no-regret learning procedure converges to the Hannan set (e.g. coarse equilibria) and that internally no-regret learning procedures lead to correlated equilibria. No-regret procedures, widely used in statistical learning, have a very large class of applications that go beyond the scope of this book. We provide the connection with calibration and refer the interested reader to [35].

The last section deals with incomplete information games (also called Bayesian games) and the natural extension of equilibria to this set-up.

7.2 Correlated Equilibria

This section is devoted to the notion of correlated equilibria, which is an extension of Nash equilibria, due to Aumann [3], and has good strategical, geometrical and dynamical properties.

© Springer Nature Switzerland AG 2019
R. Laraki et al., *Mathematical Foundations of Game Theory*, Universitext,
https://doi.org/10.1007/978-3-030-26646-2_7

7.2.1 *Examples*

Let us first consider the classical "Battle of the Sexes":

	l	r
T	3, 1	0, 0
B	0, 0	1, 3

There are two pure, efficient and disymmetrical equilibria and one mixed symmetrical and Pareto dominated equilibrium. The use of a public coin allows us to get a symmetrical and efficient outcome as an equilibrium: if the coin shows "heads", player 1 plays Top, player 2 plays left, the outcome is (3, 1) and if the coin shows "tails", player 1 plays Bottom, player 2 plays right, inducing (1, 3). Obviously facing such a plan, no deviation is profitable. This contract induces the following distribution on the action profiles:

	l	r
T	1/2	0
B	0	1/2

It is clear that a similar procedure allows us to obtain any point in the convex hull of the set of equilibrium payoffs: if $a = \sum_{r \in R} \lambda_r a_r$, $\lambda \in \Delta(R)$, R finite, use a random variable with distribution λ on R and play, if r is publicly observed, an equilibrium profile leading to the outcome a_r.

Let us consider now the following game ("Chicken"):

	l	r
T	2, 7	6, 6
B	0, 0	7, 2

Introduce a signal set (X, Y, Z) endowed with the uniform probability $(1/3, 1/3, 1/3)$. Assume that the players receive private messages on the chosen signal before playing:

1 learns $a = \{X, Y\}$ or $b = \{Z\}$;

2 learns $\alpha = \{X\}$ or $\beta = \{Y, Z\}$.

Consider the strategies:

for player 1: T if a, B if b;

for player 2: l if α, r if β.

They induce on the action space S the correlation matrix:

	l	r
T	1/3	1/3
B	0	1/3

and no deviation is profitable.

The corresponding outcome $(5, 5)$ Pareto dominates the set of symmetrical Nash outcomes.

7.2.2 Information Structures and Extended Games

We generalize now the previous construction.

Definition 7.2.1 An *information structure* $\mathcal{I}$ is defined by:

- a random event represented by a probability space $(\Omega, \mathcal{C}, P)$;
- a family of measurable maps θ^i from $(\Omega, \mathcal{C})$ to $(A^i, \mathcal{A}^i)$ (measurable set of signals of player i) (or a sub σ-algebra $\mathcal{C}^i$; in the finite case a partition of Ω).

Let G, defined by $g : S = \Pi_{i \in I} S^i \to \mathbb{R}^n$, be a strategic game. Each $S^i, i \in I$, is endowed with a σ-algebra $\mathcal{S}^i$ (in the finite case, the discrete σ-algebra).

Definition 7.2.2 The game G *extended* by $\mathcal{I}$, denoted $[G, \mathcal{I}]$, is the game, played in 2 stages:

stage 0: the random variable ω is selected according to the law P and the signal $\theta^i(\omega)$ is sent to player i;
stage 1: the players play in the game G.

A *strategy* σ^i of player i in the game $[G, \mathcal{I}]$ is a measurable map from $(A^i, \mathcal{A}^i)$ to $(S^i, \mathcal{S}^i)$ (or a measurable map from $(\Omega, \mathcal{C}^i)$ to $(S^i, \mathcal{S}^i)$).

The payoff corresponding to a profile σ is

$$\gamma[G, \mathcal{I}](\sigma) = \int_\Omega g(\sigma(\omega)) P(d\omega).$$

Once strategies and payoffs are defined, the notion of Nash equilibria in $[G, \mathcal{I}]$ is as usual but due to the specific structure of the game, one sees, in the case of a finite set of signals A^i, that σ^i is a best response to σ^{-i} if and only if, for each signal a^i having positive probability under P, the (mixed) action $\sigma^i(a^i)$ is a best response, for the payoff g, to the correlated action $\theta^{-i}[a^i]$ of the set of players $-i$, defined by

$$\theta^{-i}[a^i] = \int_\Omega \sigma^{-i}(\omega) P(d\omega|a^i).$$

This corresponds to "ex-post" rationality, given the signal, hence with an updated prior $P(d\omega|a^i)$ on the basic space Ω.

A similar property holds for general signal sets under regularity assumptions.

7.2.3   Correlated Equilibrium

The notion of correlated equilibrium extends the idea of equilibria to frameworks with informational interdependence.

Definition 7.2.3 A *correlated equilibrium* of G is a Nash equilibrium of some extended game $[G, \mathcal{I}]$.

Obviously, if the information structure is trivial one recovers Nash equilibrium.

Since Nash equilibria of the extended game correspond to profiles defined on different spaces (depending on $\mathcal{I}$), it is natural to consider the image measure on the set of action profiles S. A profile σ of strategies in $[G, \mathcal{I}]$ maps the probability P on Ω to an image probability $Q(\sigma)$ on S, which is the *correlated distribution* induced by σ: random variable $\to$ signal $\to$ action, thus: $P \to Q$.

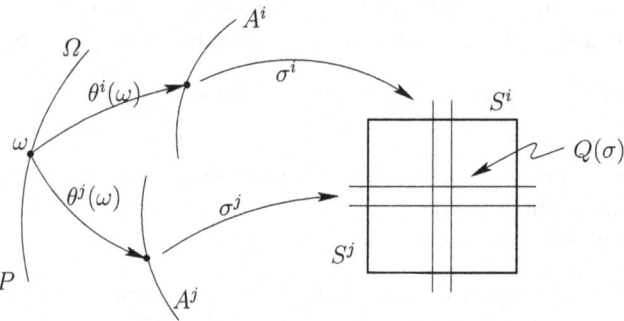

Explicitly, in the finite set-up, for each ω, let $q(\omega, \sigma)$ be the product probability on S defined by $\prod_i \sigma^i(\theta^i(\omega))$ and $Q(\sigma)$ be the expectation with respect to the underlying probability P.

The image of the equilibria profiles under this map is captured by the next definition.

Definition 7.2.4 CED(G) is the set of *correlated equilibrium distributions* in G:

$$\mathrm{CED}(G) = \bigcup_{\mathcal{I}, \sigma} \{Q(\sigma);\ \sigma \text{ equilibrium in } [G, \mathcal{I}]\}.$$

Note that CED(G) is a convex set: just consider the product of independent information structures.

7.2.4   Canonical Correlation

We introduce here a special class of information structures and equilibria that are natural and will be sufficient (in a statistical sense).

Definition 7.2.5 A *canonical information structure* $\mathcal{I}$ for G corresponds to the framework where:

- the underlying space is $\Omega = S$;
- the signal space of player i is $A^i = S^i$;
- the signaling function, $\theta^i : S \to S^i$, is defined by $\theta^i(s) = s^i$ for all $i \in I$.

Thus P is a probability on the product of action sets and once a profile of actions is chosen, each player is informed upon its own component.

A *canonical correlated equilibrium* is a Nash equilibrium of the game G extended by a canonical information structure $\mathcal{I}$ and where the equilibrium strategies are given by

$$\sigma^i(\omega) = \sigma^i(s) = \sigma^i(s^i) = s^i.$$

Thus at equilibrium, "each player plays the received signal".

Note that the associated canonical correlated equilibrium distribution (CCED) is obviously P.

The next fundamental result shows that in terms of distributions, canonical correlated equilibria form an exhaustive set.

Theorem 7.2.6 ([3])

$$\mathrm{CCED}(G) = \mathrm{CED}(G)$$

Proof Let σ be an equilibrium profile in some extension $[G, \mathcal{I}]$ and $Q = Q(\sigma)$ be the induced distribution. Let us prove that Q is also a $\mathrm{CCED}(G)$. Let $\mathcal{I}(Q)$ be the canonical information structure associated to Q. Note that each player i receives less information under $\mathcal{I}(Q)$ than under $\mathcal{I}$: only his recommended action s^i rather than the signal a^i leading to $\sigma^i(a^i) = s^i$. But s^i is a best response to the correlated strategy of $-i$ conditionally to a^i. It is then enough to use the "convexity" of BR^i on $\Delta(S^{-i})$: by linearity of player i's payoff w.r.t. $\Delta(S^{-i})$, s^i is still a best response to the correlated strategy of players $-i$ conditionally to the union of the events $\{a^i; \sigma^i(a^i) = s^i\}$, which is the event $\{s^i\}$. $\qquad\square$

7.2.5 Characterization

The previous representation theorem allows for a nice geometrical characterization of correlated equilibrium distributions. We consider the case of a finite game, but a similar property holds in general.

Theorem 7.2.7 $Q \in \mathrm{CED}(G)$ *can be written as*

$$\sum_{s^{-i} \in S^{-i}} [g^i(s^i, s^{-i}) - g^i(t^i, s^{-i})] Q(s^i, s^{-i}) \geqslant 0, \qquad \forall s^i, t^i \in S^i, \forall i \in I.$$

Proof We can assume that $Q \in \text{CCED}(G)$. If s^i is announced (i.e. its marginal probability $Q^i(s^i) = \sum_{s^{-i}} Q(s^i, s^{-i}) > 0$), we introduce the conditional distribution on S^{-i}, $Q(\cdot|s^i)$, and the equilibrium condition reads as

$$s^i \in \text{BR}^i(Q(\cdot|s^i)).$$

s^i is a best response of player i to the distribution, conditional to the signal s^i, of the actions of the other players. $\square$

Recall that the approach in terms of Nash equilibria of the extended game is an "ex-ante" analysis. The previous characterization corresponds to the "ex-post" criteria.

Corollary 7.2.8 *The set of correlated equilibrium distributions is a polytope: the convex hull of finitely many points.*

Proof $\text{CED}(G)$ is defined in $\Delta(S)$ by a finite family of weak linear inequalities. $\square$

7.2.6 Comments

An elementary existence proof of correlated equilibrium can be obtained via the minmax theorem, see Hart and Schmeidler [97] and Exercise 4.

There exist correlated equilibrium distributions outside the convex hull of Nash equilibrium distributions, or even dominating strictly in terms of payoffs. In fact, consider the game:

0, 0	5, 4	4, 5
4, 5	0, 0	5, 4
5, 4	4, 5	0, 0

The only equilibrium is symmetrical and corresponds to the strategy $(1/3, 1/3, 1/3)$ with payoff $(3,3)$.

However, the following is a correlated equilibrium distribution:

0	1/6	1/6
1/6	0	1/6
1/6	1/6	0

inducing the payoff $(9/2, 9/2)$.

Let us examine correlated equilibria in an extensive form game:

Player 1 chooses between the action "stop", inducing the payoff $(2, 2)$, or "continue", and the game is then the following:

5, 1	0, 0
0, 0	1, 5

$(3, 3)$ is an equilibrium outcome if the public signal with distribution $(1/2, 1/2)$ on (a, b) (with the convention $(5, 1)$ after a and $(1, 5)$ after b) is announced after the initial choice of player 1, but not if it is known before, since then player 1 would deviate if the signal is b.

For the study of equilibria with more general mechanisms (even incorporating a mediator) see Forges [56, 58] and Myerson [150].

7.3 No-regret Procedures

Let $\{U_n\}$ be a sequence of vectors in $\mathcal{U} = [-1, 1]^K$. At each stage n, a player having observed the previous realizations $\{U_1, \ldots, U_{n-1}\}$, and knowing his previous choices, selects a component k_n in K. The corresponding outcome is $\omega_n = U_n^{k_n}$.

A strategy in this prediction problem (guessing the largest component) specifies at each stage n the law of k_n as a function of the past history $h_{n-1} = \{k_1, U_1, \ldots, k_{n-1}, U_{n-1}\}$, denoted $\sigma(h_{n-1}) \in \Delta(K)$.

7.3.1 External Regret

The *external regret* given the choice $\ell \in K$ and the realization $U \in \mathcal{U}$ is the vector $R(\ell, U) \in \mathbb{R}^K$ defined by

$$R(\ell; U)^k = U^k - U^\ell, k \in K.$$

The evaluation at stage n is $R_n = R(k_n, U_n)$ with components $R_n^k = U_n^k - \omega_n$, hence the average external regret vector at stage n is $\overline{R}_n$ with

$$\overline{R}_n^k = \overline{U}_n^k - \overline{\omega}_n.$$

(Recall that given a sequence $\{u_m\}$, $\bar{u}_n$ denotes the average $\frac{1}{n} \sum_{m=1}^n u_m$.) It compares the actual (average) payoff to the payoff corresponding to the choice of a constant component, see Foster and Vohra [64], Fudenberg and Levine [65].

Definition 7.3.1 A strategy σ has *no external regret* (or satisfies *external consistency*) if, for every process $\{U_m\}$,

$$\max_{k \in K} \left([\overline{R}_n^k]^+ \right) \longrightarrow 0 \text{ a.s., as } n \to +\infty$$

or, equivalently,

$$\sum_{m=1}^n (U_m^k - \omega_m) \leqslant o(n), \qquad \forall k \in K.$$

To prove the existence of a strategy satisfying this property we will use approachability theory (Chap. 3, Exercise 4) and show that the negative orthant $D = \mathbb{R}^K_-$ is approachable by the sequence of regret $\{R_n\}$.

We recall Blackwell's theorem [27]:

Theorem 7.3.2 *Let $(x_n)_n$ be a bounded sequence of random variables in $\mathbb{R}^K$ such that*

$$(*) \qquad \langle \bar{x}_n - \Pi_D(\bar{x}_n), y_{n+1} - \Pi_D(\bar{x}_n) \rangle \leqslant 0,$$

where $y_{n+1} = E(x_{n+1}|x_1, \ldots, x_n)$ is the conditional expectation of x_{n+1} given the past and Π_D stands for the projection on D. Then the distance from $\bar{x}_n$ to D goes to 0 almost surely.

A crucial property is the following:

Lemma 7.3.3 $\forall x \in \Delta(K), \forall U \in \mathcal{U}$:

$$\langle x, \mathbb{E}_x[R(\cdot, U)] \rangle = 0.$$

Proof We have

$$\mathbb{E}_x[R(\cdot, U)] = \sum_{k \in K} x_k \, R(k, U) = \sum_{k \in K} x_k(U - U^k \mathbf{1}) = U - \langle x, U \rangle \mathbf{1}$$

($\mathbf{1}$ is the K-vector of ones), thus $\langle x, \mathbb{E}_x[R(\cdot, U)] \rangle = 0$. $\square$

If $\overline{R}_n^+ \neq 0$, define $\sigma(h_n) \div \overline{R}_n^+$ i.e. proportional to this vector.

Proposition 7.3.4 σ *has no external regret.*

Proof Note that D being the negative orthant, $\overline{R}_n - \Pi_D(\overline{R}_n) = \overline{R}_n^+$ and

$$\langle \Pi_D(\overline{R}_n), \overline{R}_n - \Pi_D(\overline{R}_n) \rangle = 0.$$

Hence,

$$\begin{aligned}
\langle \mathbb{E}(R_{n+1}|h_n), \overline{R}_n - \Pi_D(\overline{R}_n) \rangle &= \langle \mathbb{E}(R_{n+1}|h_n), \overline{R}_n^+ \rangle \\
&\div \langle \mathbb{E}(R_{n+1}|h_n), \sigma(h_n) \rangle \\
&= \langle \mathbb{E}_x[R(\cdot, U_{n+1})], x \rangle \\
&= 0
\end{aligned}$$

by using Lemma 7.3.3 with $x = \sigma(h_n)$.

This implies

$$\langle \mathbb{E}(R_{n+1}|h_n) - \Pi_D(\overline{R}_n), \overline{R}_n - \Pi_D(\overline{R}_n) \rangle = 0.$$

Thus condition $(*)$ is satisfied, so D is *approachable*: $d(\overline{R}_n, \mathbb{R}^K_-) \longrightarrow 0$ and $\max_{k \in K} [\overline{R}_n^k]^+ \longrightarrow 0$. $\square$

7.3.2 Internal Regret

We consider now a more precise evaluation.

The *internal regret* given $k \in K$ and $U \in \mathcal{U}$ is the $K \times K$ matrix $S(k, U)$ with

$$S(k, U)^{j\ell} = \begin{cases} U^\ell - U^k & \text{if } j = k, \\ 0 & \text{otherwise.} \end{cases}$$

The internal regret at stage n is given by $S_n = S(k_n, U_n)$ and the average regret at stage n is thus expressed by the matrix

$$\overline{S}_n^{k\ell} = \frac{1}{n} \sum_{m=1, k_m=k}^{n} (U_m^\ell - U_m^k).$$

The component $k\ell$ compares the average payoff of the player for the stages where he played k to the payoff he would have obtained by playing ℓ at all these stages, see Foster and Vohra [64] and Fudenberg and Levine [67].

Definition 7.3.5 A strategy σ has *no internal regret* (or satisfies *internal consistency*) if for each process $\{U_m\}$ and each pair k, ℓ:

$$[\overline{S}_n^{k\ell}]^+ \longrightarrow 0 \qquad \text{as } n \to +\infty \qquad \sigma\text{-p.s.}$$

Given a $K \times K$ matrix A with non-negative coefficients $\geqslant 0$, let $\chi[A]$ be the non-empty set of *invariant probability measures* for A, namely vectors $\mu \in \Delta(K)$ satisfying

$$\sum_{k \in K} \mu^k A^{k\ell} = \mu^\ell \sum_{k \in K} A^{\ell k} \qquad \forall \ell \in K.$$

(This is an easy consequence of the existence of an invariant measure for a stochastic matrix, Corollary 2.5.2.)

Lemma 7.3.6 *For each $K \times K$ positive matrix A and each measure $\mu \in \chi[A]$:*

$$\langle A, E_\mu(S(\cdot, U)) \rangle = \sum_{k, \ell} A^{k\ell} E_\mu(S(\cdot, U)^{k\ell}) = 0, \quad \forall U \in \mathcal{U}.$$

Proof

$$\langle A, E_\mu(S(\cdot, U)) \rangle = \sum_{k, \ell} A^{k\ell} \mu^k (U^\ell - U^k)$$

and the coefficient of each U^ℓ is

$$\sum_{k \in K} \mu^k A^{k\ell} - \mu^\ell \sum_{k \in K} A^{\ell k} = 0.$$

□

Define the strategy by $\sigma(h_n) = \mu(\bar{S}_n^+)$.

Proposition 7.3.7 σ *has no internal regret.*

Proof The sufficient condition for approachability of the negative orthant D of $\mathbb{R}^{K \times K}$ reads as

$$\langle \bar{S}_n - \Pi_D(\bar{S}_n), E(S_{n+1}|S_1, \dots, S_n) - \Pi_D(\bar{S}_n) \rangle \leqslant 0.$$

Here one has $\Pi_D(x) = -x^-$ orthogonal to $x^+ = x - \Pi_D(x)$, hence the condition becomes

$$\langle \bar{S}_n^+, E_P(S(\cdot, U_{n+1})) \rangle \leqslant 0,$$

which is satisfied for $P = \sigma(h_n) = \mu(\bar{S}_n^+)$, using the previous lemma with $A = \bar{S}_n^+$. Then Δ is approachable, hence $\max_{k,\ell}[\bar{S}_n^{k,\ell}]^+ \longrightarrow 0$.   □

Remarks 7.3.8 External consistency can be considered as a robustness property of σ facing a given finite family L of "external experts" using strategies $\phi^\ell, \ell \in L$:

$$\lim \frac{1}{n} \left[\sum_{m=0}^n \langle \phi_m^\ell - x_m, U_m \rangle \right]^+ = 0, \quad \forall \phi^\ell.$$

The typical case corresponds to a constant choice: $L = K$ and $\phi^k = k$.

In general, the rôle of k will be the (random) move of expert ℓ, which the player follows with probability x_m^ℓ at stage m. U_m^ℓ is then the payoff to expert ℓ at stage m and external consistency means that a player can asymptotically achieve at least the maximum of the expert payoffs.

Internal consistency corresponds to the case of experts adjusting their behavior to that of the predictor.

7.3.3 Calibrating

Consider a sequence of random variables X_m with values in a finite set Ω (which will be written as a basis of $\mathbb{R}^\Omega$).

A player observes the previous realizations X_n, $n < m$, and uses at stage m a prediction ϕ_m with values in a finite discretization V of $D = \Delta(\Omega)$. The interpretation is as follows: "$\phi_m = v$" means that the anticipated probability that $X_m = \omega$ (or $X_m^\omega = 1$) is v^ω. As usual, the player will use randomized forecasts.

Definition 7.3.9 Given $\varepsilon > 0$, a strategy ϕ is ε-*calibrated* if

$$\lim_{n \to +\infty} \frac{1}{n} \sum_{v \in V} \left\| \sum_{\{m \leqslant n, \phi_m = v\}} (X_m - v) \right\| \leqslant \varepsilon, \quad \phi \quad a.s.$$

This inequality says that if the average number of times v is predicted does not vanish, the average value of X_m over these times is close to v. More precisely, let B_n^v be the set of stages before n where v is announced, let N_n^v be its cardinality and $\overline{X}_n(v)$ be the empirical average of X_m over these stages. Then the condition reads as

$$\lim_{n \to +\infty} \sum_{v \in V} \frac{N_n^v}{n} \|\overline{X}_n(v) - v\| \leqslant \varepsilon.$$

We will use internal consistency to prove calibration, Foster and Vohra [63].

Proposition 7.3.10 *For any $\varepsilon > 0$, there exists an ε-calibrated strategy.*

Proof Consider the on-line procedure where the choice set of the forecaster is V and the outcome given v and X is

$$U^v = \|X - v\|^2,$$

where we use the Euclidean norm.

Given an internal consistent procedure σ one obtains (the outcome here is a loss)

$$\frac{1}{n} \sum_{m \in B_n^v} (U_m^v - U_m^w) \leqslant o(1), \qquad \forall w \in V,$$

which is

$$\frac{1}{n} \sum_{m \in B_n^v} (\|X_m - v\|^2 - \|X_m - w\|^2) \leqslant o(1), \qquad \forall w \in V,$$

hence implies, using the property of the Euclidean norm,

$$\frac{N_n^v}{n}(\|\overline{X}_n(v) - v\|^2 - \|\overline{X}_n(v) - w\|^2) \leqslant o(1), \qquad \forall w \in V.$$

In particular, by choosing a point w closest to $\overline{X}_n(v)$,

$$\frac{N_n^v}{n}(\|\overline{X}_n(v) - v\|^2 - \delta^2) \leqslant +o(1),$$

where δ is the L^2 mesh of V, from which calibration follows. $\qquad\square$

Remark 7.3.11 One can also use calibration to prove approachability of convex sets (Chap. 3, Exercise 4), Foster and Vohra [62]:

Assume that C satisfies $[Ay] \cap C \neq \emptyset$:

$$\forall y \in Y, \exists x \in X \text{ such that } x \, A \, y \in C.$$

Consider a δ-grid of Y defined by $\{y_v, v \in V\}$. A stage is of type v if player 1 predicts y_v and then plays a mixed move x_v such that $x_v \, A \, y_v \in C$. By using a calibrated procedure, the average of the moves of player 2 on the stages of type v will be δ close to y_v. By a martingale argument the average outcome will then be ε close to $x_v \, A \, y_v$ for δ small enough and n large enough. Finally, the total average outcome is a convex combination of such amounts, hence is close to C by convexity.

7.3.4 Application to Games

Consider a finite game in strategic form: $g : S = \prod_i S^i \longrightarrow \mathbb{R}^I$. Since the analysis will be done from the view point of player 1 it is convenient to set $S^1 = K$, $X = \Delta(K)$ (mixed moves of player 1), $L = \prod_{i \neq 1} S^i$, and $Y = \Delta(L)$ (correlated mixed moves of player 1's opponents), hence if $Z = \Delta(S)$ (correlated moves), one has $Z = \Delta(K \times L)$.

$F = g^1$ and we still denote by F its linear extension to Z, and its bilinear extension to $X \times Y$. The repeated game where the moves are announced corresponds to the on-line process with $\{U_n^k = F(k, \ell_n)\}$ being the vector payoff for player 1 induced by the actions at stage n of his opponents.

7.3.4.1 External Consistency and Hannan's Set

Given $z \in Z$, let
$$r(z) = \{F(k, z^{-1}) - F(z)\}_{k \in K},$$

where z^{-1} stands for the marginal of z on L. (Player 1 compares his payoff using a given move k to his payoff under z, assuming the other players' behavior is given.) Let $D = \mathbb{R}_-^K$ be the closed negative orthant associated to the set of moves of player 1.

Definition 7.3.12 ([89]) H (for Hannan's set) is the set of correlated moves satisfying the no r-regret condition for player 1:

$$H = \{z \in Z : F(k, z^{-1}) \leqslant F(z), \forall k \in K\} = \{z \in Z : r(z) \in D\}.$$

Define the empirical average distribution:

$$z_n = \frac{1}{n} \sum_{m=1}^{n} (k_m, \ell_m) \in Z.$$

Proposition 7.3.13 *If player 1 follows a strategy with no external regret, the empirical distribution of moves converges a.s. to the Hannan set.*

Proof The proof is straightforward due to the linearity of the payoff. The no-regret property reads as

$$\frac{1}{n}\sum_{m=1}^{n} F(k, \ell_m) - \frac{1}{n}\sum_{m=1}^{n} F(k_m, \ell_m) \leqslant o(1) \qquad \forall k \in K,$$

which gives

$$F(k, \tfrac{1}{n}\sum_{m=1}^{n} \ell_m)) - F(\tfrac{1}{n}\sum_{m=1}^{n}(k_m, \ell_m)) \leqslant o(1) \qquad \forall k \in K,$$

and this expression is

$$F(k, z_n^{-1}) - F(z_n) \leqslant o(1) \qquad \forall k \in K.$$

□

In the framework of two players one can define Hannan sets (H^i) for each and consider their intersection $\overline{H}$. In particular, in the case of a zero-sum game one has, for $z \in \overline{H}$,

$$F(z) \geqslant F(z_1, z_2)$$

and the opposite inequality for the other player, hence the marginals are optimal strategies and $F(z)$ is equal to the value.

Example 7.3.14 For the game:

0	1	−1
−1	0	1
1	−1	0

consider the distribution:

1/3	0	0
0	1/3	0
0	0	1/3

7.3.4.2 Internal Consistency and Correlated Equilibria

We still consider only player 1 and denote by F his payoff. Given $z = (z_s)_{s \in S} \in Z$, introduce the family of m comparison vectors of dimension m (testing k against j with $(j, k) \in K^2$) defined by

$$C(j, k)(z) = \sum_{\ell \in L} [F(k, \ell) - F(j, \ell)]z_{(j,\ell)}.$$

(This corresponds to the change in the expected gain of player 1 at z when replacing move j by k.) Observe that if one denotes by $(z \mid j)$ the conditional probability on L induced by z given $j \in K$ and by z^1 the marginal on K, then

$$\{C(j,k)(z)\}_{k \in K} = z_j^1 r((z \mid j)),$$

where we recall that $r((z \mid j))$ is the vector of regrets for player 1 at $(z \mid j)$.

Definition 7.3.15 The set of *no C-regret* (for player 1) is

$$C^1 = \{z; C(j,k)(z) \leqslant 0, \forall j, k \in K\}.$$

It is obviously a subset of H since

$$\sum_j \{C(j,k)(z)\}_{k \in I} = r(z).$$

As above, when considering the payoff vector generated by the actions of the opponents in the repeated game, we obtain:

Proposition 7.3.16 *If player 1 follows some no internal regret strategy, the empirical distribution of moves converges a.s. to the set C^1.*

Proof The no internal regret property is

$$\frac{1}{n} \sum_{m=1, j_m=j}^{n} (F(k, \ell_m) - F(j, \ell_m)) \leqslant o(1).$$

Hence,

$$\sum_\ell z_n(j, \ell)(F(k, \ell) - F(j, \ell)) \leqslant o(1).$$

□

Recall that the set of correlated equilibrium distributions of the game with payoff g is defined by (Theorem 7.2.7)

$$C = \{z \in Z; \sum_{\ell \in L} [g^i(k, \ell) - g^i(j, \ell)] z_{(j, \ell)} \leqslant 0, \qquad \forall j, k \in S^i, \forall i \in I\}.$$

This implies:

Proposition 7.3.17 *The intersection over all players i of the sets C^i is the set of correlated equilibrium distributions of the game.*

Hence, in particular:

Proposition 7.3.18 *If each player follows some no internal regret procedure, the distance from the empirical distribution of moves to* CED(G) *converges a.s. to 0.*

This important result implies the existence of CCED and gives an algorithm (generated by unilateral procedures) converging to this set. There are no such properties for Nash equilibria. For a survey on this topic, see Hart [95].

7.4 Games with Incomplete Information (or Bayesian Games)

7.4.1 Strategies, Payoffs and Equilibria

As Sect. 7.2 above, an information structure $\mathcal{I} = (\Omega, \mathcal{A}, P)$ is given, but the game itself $\Gamma = G(\cdot)$ has a payoff which is a function of the random variable ω, called the state.

It is usual to call the (finite) set of signals A^i the *type set* (each player knows his type). A strategy σ^i of player i is a map from A^i to $\Delta(S^i)$. The (vector) payoff corresponding to a profile σ is given by

$$\gamma(\sigma) = \int_\omega g(\{\sigma^i(\theta^i(\omega))\}_{i \in I}; \omega) P(d\omega).$$

Denote by Π the induced probability on the product type set $A = \Pi_i A^i$ and $g(\{\sigma^j(a^j)\}, a)$ the corresponding conditional expectation of $g(\{\sigma^j(a^j)\}, \omega)$ on $\theta^{-1}(a)$, which is the set of random states inducing the profile of signals a. Then the payoff can also be written as

$$\gamma(\sigma) = \sum_a g(\{\sigma^j(a^j)\}; a)\Pi(a),$$

hence for player i,

$$\gamma^i(\sigma) = \sum_{a^i} \Pi^i(a^i) B^i(a^i),$$

where Π^i is the marginal of Π on A^i and

$$B^i(a^i) = \sum_{a^{-i}} g^i(\sigma^i(a^i), \{\sigma^j(a^j)\}_{j \neq i}; (a^i, a^{-i}))\Pi(a^{-i}|a^i).$$

Hence if σ is an equilibrium profile, for each player i and each signal a^i, $\sigma^i(a^i)$ maximizes the "Bayesian" payoff facing σ^{-i}, which is $B^i(a^i)$.

The first maximization (in γ) is "ex-ante", and the second one "ex-post".

7.4.2 Complements

A *pure* (resp. *behavioral*) strategy of player i maps A^i to S^i (resp. $\Delta(S^i)$).

A *mixed strategy* is a probability on the set of pure strategies or a measurable map from $A^i \times [0, 1]$ to S^i, where $[0, 1]$ is endowed with the uniform measure.

A *distributional strategy* μ^i is an element of $\Delta(A^i \times S^i)$ which is compatible with the data: the marginal distribution on A^i is Π^i [141].

The conditional probability $\mu^i(\cdot|a^i)$ corresponds to a behavioral strategy.

Example 7.4.1 (War of attrition [26])
Consider a symmetric two-player game: each player $i = 1, 2$ has a type $a^i \in A^i = \mathbb{R}^+$ and chooses an action $s^i \in S^i = \mathbb{R}^+$. The payoff of each player is a function of his own type and of both actions:

$$f^i(a^i; s^1, s^2) = \begin{cases} a^i - s^j & \text{if } s^j < s^i \\ -s^i & \text{otherwise.} \end{cases}$$

The distribution of the types are independent and given by a cumulative distribution function G, known to the players. In addition, each player knows his type.

One interpretation is a game of timing where the last remaining player wins a function of his type and of the duration of the conflict while the other loses his time.

1. If G corresponds to the Dirac mass at some point v, the only symmetric equilibrium is to play s according to the law Q_v, where

$$\mathbb{P}(s \leqslant t) = 1 - \exp\left(-\frac{t}{v}\right).$$

2. If G has a density g, the only symmetric equilibrium is to play, given the type a^i, the (pure) action

$$s(a^i) = \int_0^{a^i} \frac{t\, g(t)}{1 - G(t)} dt.$$

3. Let G_n with density g_n on $[v - \frac{1}{n}, v + \frac{1}{n}]$ weak* converge to the Dirac mass at v. At equilibrium the distribution of actions of i is given by the following:

$$\Phi_n(t) = \mathbb{P}(s \leqslant t) = P_{G_n}(a^i ; s(a^i) \leqslant t).$$

Then Φ_n converges to Q_v.

There is a natural link with the following model of non-atomic games [128, 186], where the natural variable will be defined on a product space.

The framework is defined by:

– a set of agents A, endowed with a non-atomic measure μ;

and for each agent:

– an action space S;
– a payoff F, as a function of his own type a, his own action s, and the distribution ν of the actions of the other players.

An equilibrium is thus a measure λ on $A \times S$, with marginal μ on A and ν on S satisfying:

$$\lambda\{(a, s) \in A \times S; s \in \operatorname{argmax} F(a, \cdot, \nu)\} = 1.$$

The strategic use of information is a fundamental field with several active directions of research such as:

– private information and purification of strategies [92];
– reputation phenomena: interactive repeated situations where one player uses in a strategic way the uncertainty of his opponents on his type to build a controlled process of beliefs, see e.g. Aumann and Sorin [15], Sorin [202] and the relation with repeated games (Chap. 8), see Mailath and Samuelson [124].

7.5 Exercises

Exercise 1. Value of information (Kamien, Taumann and Zamir [105])
Consider the following two-player game in extensive form:

Stage 0: A color (white or black, W or B) is chosen by Nature uniformly.
Stage 1: Player 1 announces a color in $\{W, B\}$ and this is communicated to player 2.
Stage 2: Player 2 announces a color in $\{W, B\}$.

The payoff is 2 for both players if they announce the same color, otherwise, it is 5 for the player who predicted the right color and 0 to the other player. We consider several informational variants of this game, and in all of them the description is common knowledge among the players (for example, if player 1 has private information, player 2 knows this fact, and player 1 knows that player 2 knows, etc).

(1) Suppose that no player observes Nature's move. Write the game in extensive and normal forms and compute the unique equilibrium.
(2) Suppose that player 1 only observes Nature's move. Write the game in extensive and normal forms and compute the unique equilibrium.
(3) Finally, solve in a similar manner the cases where only player 2 observes Nature's move and where both players observe Nature's move. Conclude.

Exercise 2. Strategic transmission of information
Consider a two-player strategic interaction in which a state of Nature $k \in \{1, 2\}$ is chosen according to the uniform distribution. The state is then observed by player 1

(but not player 2). Player 1 then sends a message $m \in \{A, B\}$ to player 2. Then player 2 chooses an action $a \in \{L, M, R\}$. The payoff depends only on the pair state/action (k, a) but not on the message m:

State $k = 1$: Payoffs for each action are: $L \to (0,6)$; $M \to (2,5)$; $R \to (0,0)$.
State $k = 2$: Payoffs for each action are: $L \to (0,0)$; $M \to (2,5)$; $R \to (2,12)$.

(1) Write the game in extensive and normal form.
(2) Compute all Nash equilibria in pure strategies and their corresponding payoffs, classified by the number of different messages (one or two) sent by player 1 on the equilibrium path. (When only one message is sent, the equilibrium is called non-revealing or pooling, and when both messages are sent the equilibrium is fully revealing.)
(3) Show that the following profile in behavioral strategies is a Nash equilibrium and compute its associated payoff:

Player 1: plays A if $k = 1$ and $\frac{1}{2}A + \frac{1}{2}B$ if $k = 2$;
Player 2: plays M if he observes A, and plays R if he observes B.

This equilibrium is partially revealing.
(4) Conclude.

Exercise 3. Correlated equilibria versus Nash: existence (Peleg [162])

(1) Show that the following game:

	b_1	b_2	b_3
a_1	$(-\infty, -\infty)$	$(3, 1)$	$(0, 2)$
a_2	$(1, 3)$	$(0, 0)$	$(1, -\infty)$
a_3	$(2, 0)$	$(-\infty, 1)$	$(0, 0)$

has no Nash equilibrium but any distribution of the form

	b_1	b_2	b_3
a_1	0	α	0
a_2	β	γ	0
a_3	0	0	0

with $\alpha\beta\gamma > 0$ is a correlated equilibrium distribution.

(2) Consider the game with an infinite countable set of players $\mathbb{N}^* = \{1, 2, 3, \dots\}$. Assume that each player has two strategies 0 or 1 (thus $S^i = \{0, 1\}$). The payoff function of player i is

$$g^i(s) = \begin{cases} s^i, & \text{if } \sum_j s^j < \infty \\ -s^i, & \text{otherwise.} \end{cases}$$

(a) Show that there exists no Nash equilibrium in pure strategies.
(b) Use the Borel–Cantelli Lemma to prove that there are no mixed equilibria either.
(c) Prove that the distribution $\mu = \frac{\mu_1}{2} + \frac{\mu_2}{2}$ on $S = \prod_i S^i = \{0, 1\}^{\mathbb{N}^*}$ induces a correlated equilibrium where:

μ_1 is the product distribution $\mu_1 = \bigotimes_i \mu_1^i$ with: $\mu_1^i(s^i = 1) = \frac{1}{i}$;

μ_2 is the joint distribution where the profile $(s^1 = 1, \ldots, s^i = 1, s^{i+1} = 0, \ldots, s^n = 0, \ldots)$ has probability $\frac{1}{i} - \frac{1}{i+1} = \frac{1}{i(i+1)}$. (Note that $\mathbb{P}_{\mu_1}(\sum s^i = \infty) = 1$, $\mathbb{P}_{\mu_2}(\sum s^i = \infty) = 0$ and $\mathbb{P}_{\mu_2}(s_i = 1) = \frac{1}{i}$.)

Exercise 4. Correlated equilibrium distribution via minmax (Hart and Schmeidler [97])

Let G be a strategic two-player game with strategy sets S^1 and S^2 and payoff $g : S = S^1 \times S^2 \longrightarrow \mathbb{R}^2$. Consider now the game Γ which is two-player and zero-sum with strategy sets S and $L = (S^1)^2 \cup (S^2)^2$ and payoff γ defined by

$$\gamma(s; t^i, u^i) = (g^i(t^i, s^{-i}) - g^i(u^i, s^{-i}))1_{\{t^i = s^i\}}.$$

(1) Verify that Γ has a value v and optimal strategies.
(2) Show that if $v \geqslant 0$ and $Q \in \Delta(S)$ is an optimal strategy of player 1, then Q is a correlated equilibrium distribution in G.
(3) Let $\pi \in \Delta(L)$. Define ρ^1, a transition probability on S^1, by

$$\rho^1(t^1; u^1) = \pi(t^1, u^1), \quad \text{if } t^1 \neq u^1,$$
$$\rho^1(t^1; t^1) = 1 - \sum_{u^1 \neq t^1} \pi(t^1, u^1).$$

Let now μ^1 be a probability on S^1 invariant under ρ^1:

$$\mu^1(t^1) = \sum_{u^1} \mu^1(u^1)\rho(u^1; t^1).$$

Define ρ^2 and μ^2 similarly and let $\mu = \mu^1 \times \mu^2$.
Show that the payoff $\gamma(\mu; \pi)$ can be decomposed into terms of the form

$$\sum_{t^1} \mu^1(t^1) \sum_{u^1} \rho(t^1; u^1)(g^1(t^1, \cdot) - g^1(u^1, \cdot))$$

and then deduce that

$$\forall \pi \in \Delta(L), \exists \phi \in \Delta(S) \text{ satisfying } \gamma(\phi, \pi) \geqslant 0.$$

(4) Prove the existence of a correlated equilibrium distribution in G.
(5) Extend the proof to the case of I players.

Exercise 5. Correlated equilibrium: zero-sum case

(1) Consider a finite two-person zero-sum game defined by $g : S = S^1 \times S^2 \to \mathbb{R}$. Given $\pi \in \Delta(S)$, let $g(\pi) = \sum_{s \in S} \pi(s^1, s^2) g(s^1, s^2)$.

   (a) Show that the only correlated equilibrium payoff $g(\pi)$ is $v = \text{val} g$.
   (b) Let $\pi \in \text{CED}(g)$ and $s^1 \in S^1$ have positive probability under π. Show that the conditional probability $\pi(\cdot | s^1) \in \Delta(S^2)$ is an optimal strategy of player 2.

(2) Consider the following zero-sum game [57]:

	b_1	b_2	b_3
a_1	0	0	1
a_2	0	0	-1
a_3	-1	1	0

   (a) Show that

	b_1	b_2	b_3
a_1	1/3	1/3	0
a_2	1/3	0	0
a_3	0	0	0

   is a correlated equilibrium distribution. Note that it is not a product of optimal strategies of the players.
   (b) Describe the set $\text{CED}(g)$.

Exercise 6. Correlated and Nash equilibria: comparison

(1) Describe all Nash and correlated equilibria of the following game:

	L	R
T	$(2, 2)$	$(0, 0)$
B	$(0, 0)$	$(2, 2)$

(2) Consider the following three-player game:
   Player 1 has two actions: T or B. Similarly player 2 has two actions: L or R. Player 3 has infinitely many actions: he chooses an integer $z \in \mathbb{Z}$. The payoff is as follows:

	L	R
T	$(2, 2, 3 + 1/z)$	$(0, 0, 8)$
B	$(0, 0, 0)$	$(2, 2, 0)$

$z < 0$

	L	R
T	$(2, 2, 2)$	$(0, 0, 0)$
B	$(0, 0, 0)$	$(2, 2, 2)$

$z = 0$

	L	R
T	$(2, 2, 0)$	$(0, 0, 0)$
B	$(0, 0, 8)$	$(2, 2, 3 - 1/z)$

$z > 0$

As usual player 1 chooses the line, player 2 the column and player 3 the matrix. (If player 1 chooses T, player 2 chooses L and player 3 chooses $z = -4$, players 1 and 2 get 2 and player 3 obtains 11/4.)

(a) Show that this game has no equilibrium.
(b) Show that there exists a correlated equilibrium where player 3 plays $z = 0$.

7.6 Comments

The notion of correlated equilibrium is an expression of Bayesian rationality when the players have access to some random information (Aumann).

It proved very successful for its robust properties: geometrical (polytope), simple proof of existence, stability by public lotteries...

The no-regret procedure provides an elegant decentralized and simple algorithm, which also gives an alternative proof of existence. No similar result can be established for Nash equilibrium [96].

An important literature on learning deals with the case where the player is not aware of the vector of outcomes that could occur, corresponding to all his possible actions. He could know only the actual realized outcome (the bandit framework) or more generally a signalling structure depending on all actions could be given.

All these approaches in terms of regret are very much in the spirit of robust statistical procedures and go back to the initial point of view of Blackwell and Girshick.

Note, however, that the framework is more "best response" than "sure strategy" since the idea is to adjust to (and eventually take advantage of) the behavior of the opponent (or to "learn" the environment). Obviously, in the zero-sum case, optimality is achieved.

Chapter 8
Introduction to Repeated Games

8.1 Introduction

Repeated games represent dynamic interactions in discrete time. In the most general setup, these interactions are modeled by a Markovian state variable which is jointly controlled by the players. The game is played in stages, and each player first receives a private signal at the initial state. Then at every stage, the players simultaneously choose an action in their own action set. The selected actions together with the state determine: (a) the current payoffs, and (b) a probability distribution over the next state and the signals received by the players. This is a very general model, and when a player selects an action at some stage, several strategic aspects are present: (1) he may influence his current payoff, (2) he may influence the state process (this aspect is crucial in the class of *stochastic games*), (3) he may reveal or learn something about the current state (this aspect is crucial in the class of *repeated games with incomplete information*), and (4) he may influence the knowledge of all players on the current action profile played at that stage (this aspect is crucial in the class of *repeated games with imperfect observation*).

Except in the last section, we only consider here the simplest case of repeated games: non-stochastic repeated games with complete information and perfect observation, which we call *standard repeated games*. The same stage game is repeated over and over, all players know the stage game and after every stage the actions played are publicly observed. We introduce the finitely repeated game where players play finitely many stages and want to maximize their average payoffs, the discounted game where players play infinitely many stages and want to maximize their discounted sum of payoffs, and the uniform game where players play infinitely many stages and want to play well in any long enough game (or in any discounted game with low enough discount factor). Repetition here opens the door to new phenomena, and players can sustain a cooperative path by threats of punishment if one player deviates from the path. This leads to a huge increase of the set of equilibrium payoffs incarnated by the famous Folk theorems: in particular, the equilibrium payoffs of the uniform game are precisely the payoffs which are feasible (achievable) and individually rational (where

© Springer Nature Switzerland AG 2019
R. Laraki et al., *Mathematical Foundations of Game Theory*, Universitext,
https://doi.org/10.1007/978-3-030-26646-2_8

each player gets at least his punishment level). The message is simple: if players are patient enough, any reasonable payoff can be achieved at equilibrium. Folk theorems for discounted games and for finitely repeated games, for Nash equilibria and for subgame-perfect equilibria, are also presented.

The last section presents extensions of the model in three directions:

(1) Repeated games with signals (or imperfect monitoring), where players imperfectly observe the actions played at the end of each stage. The Folk theorem may fail, and computing the equilibrium payoffs is challenging.
(2) Stochastic games, where the state variable which determines the current payoff function evolves from stage to stage according to the action profiles played. A particular example, the celebrated "Big Match", is studied in detail.
(3) Repeated games with incomplete information, where the stage game is fixed but imperfectly known to the players. We consider the case of two-player zero-sum games with lack of information on one side (player 1 perfectly knows the stage game whereas player 2 only has an a priori on this game) and prove the famous cav u theorem of Aumann and Maschler, characterizing the limit value of the repeated game.

8.2 Examples

Given a finite strategic game G called the "stage game", we define G_T as the game G repeated T times, the payoff of the players being defined by the arithmetic mean of their stage payoffs. G_T is a finite extensive-form game, and we denote by E_T its set of mixed Nash equilibrium payoffs.

Example 8.2.1 The stage game is

	L	R
T	$(1, 0)$	$(0, 0)$
B	$(0, 0)$	$(0, 1)$

$(1, 0)$ and $(0, 1)$ are Nash equilibrium payoffs of the stage game. It is easy to construct a Nash equilibrium of the 2-stage game with payoff $(1/2, 1/2)$: the two players play (T, L) at stage 1, and (B, R) at stage 2.

Consequently $(1/2, 1/2) \in E_2$.

Repetition allows for the convexification of the equilibrium payoffs.

Example 8.2.2 The stage game is

	C^2	D^2	E^2
C^1	$(3, 3)$	$(0, 4)$	$(-10, -10)$
D^1	$(4, 0)$	$(1, 1)$	$(-10, -10)$
E^1	$(-10, -10)$	$(-10, -10)$	$(-10, -10)$

The set of Nash equilibrium payoffs of the stage game is

$$E_1 = \{(1, 1), (-10, -10)\}.$$

One can construct a Nash equilibrium of the 2-stage game with payoff $(2, 2)$ as follows.

In the first stage, player 1 plays C^1 and player 2 plays C^2. At the second stage, player 1 plays D^1 if player 2 has played C^2 at stage 1, and he plays E^1 (this can be interpreted as a *punishment*) otherwise. Similarly, player 2 plays D^2 if player 1 has played C^1 at stage 1, and he plays E^2 (punishment) otherwise. We have defined a Nash equilibrium of the 2-stage game with payoff $(2, 2)$.

Similarly, one can show that for each $T \geqslant 1$, we have $\frac{T-1}{T}(3, 3) + \frac{1}{T}(1, 1) \in E_T$. *Using punishments, repetition may allow for cooperation.*

Example 8.2.3 The stage game is the following prisoner's dilemma:

	C^2	D^2
C^1	$(3, 3)$	$(0, 4)$
D^1	$(4, 0)$	$(1, 1)$

We show by induction that $E_T = \{(1, 1)\}$ for each T.

Proof The result is clear for $T = 1$. Assume it is true for a fixed $T \geqslant 1$, and consider a Nash equilibrium $\sigma = (\sigma^1, \sigma^2)$ of the $(T + 1)$-stage repeated game. Denote by x, respectively y, the probability that at the first stage player 1 plays C^1, respectively that player 2 plays C^2. After every action profile played with positive probability at stage 1, the continuation strategies induced by σ form a Nash equilibrium of the remaining game. Hence by assumption the average payoff associated to these continuation strategies is $(1, 1)$. Hence the equilibrium payoff for player 1 in the $(T + 1)$-stage game is $\frac{1}{T+1}((3xy + 4(1 - x)y + (1 - x)(1 - y)) + T)$. By playing D^1 at all stages, player 1 can make sure his expected payoff will be at least: $\frac{1}{T+1}((4y + (1 - y)) + T)$. By the equilibrium property, we have: $3xy + 4(1 - x)y + (1 - x)(1 - y) \geqslant 4y + (1 - y)$, and we obtain $x = 0$. Similarly $y = 0$, and the equilibrium payoff in the $T + 1$-stage game is $(1, 1)$. □

Hence there is no possible cooperation in the standard version of the prisoner's dilemma repeated a finite number of times.

8.3 A Model of Standard Repeated Games

We fix a finite strategic game $G = (N, (A^i)_{i \in N}, (g^i)_{i \in N})$, called the stage game. We will study the repetition of G a large number of times, and the infinite repetition of G. At every stage the players simultaneously choose an action in their own set of actions, then the action profile is publicly observed and the next stage is played. As usual we put $A = \prod_{i \in N} A^i$ and $g = (g^i)_{i \in N}$.

8.3.1 Histories and Plays

A history of length t is defined as a vector $(a_1, \ldots, a_t)$ of elements of A, with a_1 representing the action profile played at stage 1, a_2 representing the action profile played at stage 2, etc. The set of histories of length t is the cartesian product $A^t = A \times \cdots \times A$ (t times), denoted by H_t. (For $t = 0$ there is a unique history of length 0, and by convention H_0 is the singleton $\{\varnothing\}$.)

$$H_T = \{(a_1, \ldots, a_T), \text{ for all } t \quad a_t \in A\}.$$

The set of all histories is $H = \bigcup_{t \geqslant 0} H_t$. A *play* of the repeated game is defined as an infinite sequence $(a_1, \ldots, a_t, \ldots)$ of elements of A, the set of plays is denoted by H_∞ (identical to the cartesian product A^∞).

8.3.2 Strategies

We define a single notion of strategies. This notion corresponds to behavior strategies, and is adapted to any number of repetitions of G.

Definition 8.3.1 A *strategy* of player i is a mapping σ^i from H to $\Delta(A^i)$. We denote by Σ^i the set of strategies of player i and by $\Sigma = \prod_{i \in N} \Sigma^i$ the set of strategy profiles.

The interpretation of the strategy σ^i is the following: for each h in H_t, $\sigma^i(h)$ is the probability used by player i to select his action of stage $t + 1$ if history h has been played at the first t stages.

At every stage, given the past history, the lotteries used by the players are independent, and a strategy profile σ naturally induces by induction a probability distribution on the set H (which is countable): first use $((\sigma_1^i)_i)$ to define the probability induced by σ on the actions of stage 1, then use the transitions given by $((\sigma_2^i)_i)$ to define the probability induced by σ on the actions of stages 1 and 2, etc. Using the Kolmogorov (or Carathéodory) extension theorem, this probability can be extended in a unique way to the set of plays H_∞, endowed with the product σ-algebra on A^∞ (just as tossing a coin at every stage induces a probability distribution over sequences of Heads and Tails). Notice also that the players have perfect recall, and an extension of Kuhn's theorem applies [2].

8.3.3 Payoffs

In a repeated game the players receive a payoff at every stage, how should they evaluate their stream of payoffs? There are several possibilities, and we will consider: finitely repeated games with average payoffs, infinitely repeated discounted games, and uniform games (which are infinitely repeated and undiscounted games). In the sequel a_t denotes the random variable of the action profile played at stage t.

Definition 8.3.2 (*The finitely repeated game G_T*) The *average payoff* of player i up to stage T, if the strategy σ is played, is

$$\gamma_T^i(\sigma) = \mathbb{E}_\sigma\left(\frac{1}{T}\sum_{t=1}^{T} g^i(a_t)\right).$$

For $T \geqslant 1$, the T-*stage repeated game* is the strategic game $G_T = (N, (\Sigma^i)_{i \in N}, (\gamma_T^i)_{i \in N})$.

In the game G_T it is harmless to restrict the players to use strategies for the first T stages only, hence G_T can be seen as a finite game and by Nash's theorem its set of Nash equilibrium payoffs E_T is non-empty and compact. Notice that considering the sum of the payoffs instead of the average would not change the Nash equilibria of G_T.

Definition 8.3.3 (*The discounted game G_λ*) Given λ in $(0, 1]$, the *repeated game with discount rate λ* is $G_\lambda = (N, (\Sigma^i)_{i \in N}, (\gamma_\lambda^i)_{i \in N})$, where for each strategy profile σ:

$$\gamma_\lambda^i(\sigma) = \mathbb{E}_\sigma\left(\lambda \sum_{t=1}^{\infty} (1 - \lambda)^{t-1} g^i(a_t)\right).$$

With this definition, receiving a payoff of $1 - \lambda$ today is equivalent to receiving a payoff of 1 tomorrow. It is also standard to introduce the *discount factor* $\delta \in [0, 1)$, and the *interest rate* $r > 0$ such that $\delta = 1 - \lambda = \frac{1}{1+r}$. Consider now the discounted game where the players are restricted to play pure strategies. Using the product topology on strategy spaces ($\sigma_t^i \xrightarrow[t \to \infty]{} \sigma^i$ iff for each h in H, $\sigma_t^i(h) \xrightarrow[t \to \infty]{} \sigma^i(h)$), Theorem 4.7.3 shows that the set of mixed Nash equilibrium payoffs of this game is non-empty and compact. Since mixed and behavior strategies are equivalent, we obtain that the set E_λ of Nash equilibrium payoffs of G_λ is non-empty and compact.

Notice that both choosing $T = 1$ in Definition 8.3.2 or $\lambda = 1$ in Definition 8.3.3 leads to the consideration of the stage game $G = G_1$.

The uniform approach directly considers long-term strategic aspects.

Definition 8.3.4 (*The uniform game G_∞*) A strategy profile σ is a *uniform equilibrium* of G_∞ if:

(1) $\forall \varepsilon > 0$, σ is a ε-Nash equilibrium of any long enough finitely repeated game, i.e.: $\exists T_0, \forall T \geqslant T_0, \forall i \in N, \forall \tau^i \in \Sigma^i, \ \gamma_T^i(\tau^i, \sigma^{-i}) \leqslant \gamma_T^i(\sigma) + \varepsilon$; and

(2) $\left((\gamma_T^i(\sigma))_{i\in N}\right)_T$ has a limit $\gamma(\sigma)$ in $\mathbb{R}^N$ as T goes to infinity.

$\gamma(\sigma)$ is then called a *uniform equilibrium payoff* of the repeated game, and the set of uniform equilibrium payoffs is denoted by E_∞.

A uniform equilibrium is robust to the number of stages (which may be unknown to the players) as soon as it is large enough. Moreover, it is also robust to discount factors (which may not be the same for each player), as soon as they are low enough: the analogs of conditions (1) and (2) for discounted games also hold.

Lemma 8.3.5 *Let σ be a uniform equilibrium of G_∞. Then:*

(1') $\forall \varepsilon > 0$, σ *is a ε-Nash equilibrium of any discounted repeated game with low enough discount factor, i.e.: $\exists \lambda_0 \in (0, 1]$, $\forall i \in N$, $\forall \lambda \leqslant \lambda_0$, $\forall \tau^i \in \Sigma^i$,*
$\gamma_\lambda^i(\tau^i, \sigma^{-i}) \leqslant \gamma_\lambda^i(\sigma) + \varepsilon$; *and*
(2') $\left((\gamma_\lambda^i(\sigma))_{i\in N}\right)_\lambda$ *also converges to $\gamma(\sigma) = \lim_{T\to\infty}(\gamma_T^i(\sigma))_{i\in N}$ as λ goes to 0.*

The proof uses the following property. Consider a bounded sequence $(x_t)_{t\geqslant 1}$ of real numbers, and let $\bar{x}_t$ be $\frac{1}{t}\sum_{s=1}^{t} x_s$ for each $t \geqslant 1$. For $\lambda \in (0, 1]$, a simple computation shows that the λ-discounted sum (Abel sum) of the sequence $(x_t)_{t\geqslant 1}$ can be written as a weighted average of the Cesàro means $(\bar{x}_t)_t$:

$$\lambda \sum_{t=1}^{\infty} (1-\lambda)^{t-1} x_t = \lambda^2 \sum_{t=1}^{\infty} t(1-\lambda)^{t-1}\bar{x}_t. \tag{8.1}$$

For a proof of Lemma 8.3.5 based on (8.1), see Exercise 3 at the end of the chapter.

Proposition 8.3.6 *For each $T \geqslant 1$ and $\lambda \in (0, 1]$, we have $E_1 \subset E_T \subset E_\infty$ and $E_1 \subset E_\lambda \subset E_\infty$.*

The proof that $E_1 \subset E_T \subset E_\infty$ is simple and can be made by concatenation of equilibrium strategies. The proof that $E_\lambda \subset E_\infty$ is more subtle and can be deduced from the Folk Theorem 8.5.1.

8.4 Feasible and Individually Rational Payoffs

The set of vector payoffs achievable with correlated strategies in the stage game is $g(\Delta(A)) = \{g(P), P \in \Delta(A)\}$. Notice that it is the convex hull of the vector payoffs achievable with pure strategies.

Definition 8.4.1 The set of *feasible payoffs* is co $g(A) = g(\Delta(A))$.

The set of feasible payoffs is a bounded polytope, which represents the set of payoffs that can be obtained in any version of the repeated game. It contains E_T, E_λ and E_∞.

Definition 8.4.2 For each player i in N, the *punishment level of player i* or *threat point* is

$$v^i = \min_{x^{-i} \in \prod_{j \neq i} \Delta(A^j)} \max_{x^i \in \Delta(A^i)} g^i(x^i, x^{-i}).$$

v^i is called the *independent minmax* of player i. Be careful that in general one cannot exchange $\min_{x^{-i} \in \prod_{j \neq i} \Delta(A^j)}$ and $\max_{x^i \in \Delta(A^i)}$ in the above expression, see Exercise 2 at the end of Chap. 2.

Definition 8.4.3 The set of *individually rational payoffs* is

$$\text{IR} = \{u = (u^i)_{i \in N}, u^i \geqslant v^i \ \forall i \in N\}$$

and the set of feasible and individually rational payoffs is

$$E = (\text{co } g(A)) \cap \text{IR}.$$

Using the fact that actions are publicly observed after each stage, given a strategy profile σ^{-i} of the players different from i in Σ^{-i}, it is easy to construct a strategy σ^i of player i such that $\forall T$, $\gamma_T^i(\sigma^i, \sigma^{-i}) \geqslant v^i$. As a consequence E_∞, E_T and E_λ are always included in E.

We now illustrate the previous definitions on the prisoner's dilemma:

	C^2	D^2
C^1	(3, 3)	(0, 4)
D^1	(4, 0)	(1, 1)

We have $v^1 = v^2 = 1$, and the set of feasible and individually rational payoffs is represented in the following picture:

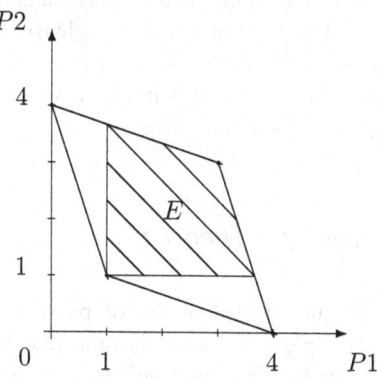

8.5 The Folk Theorems

These theorems deal with very patient players, and more precisely with: finitely repeated games with a large number of stages, discounted repeated games with a low discount factor λ, or uniform games. The message is essentially the following: *the set of equilibrium payoffs of the repeated game is the set of feasible payoffs (i.e. that one can obtain while playing) which are individually rational (i.e. such that each player gets at least his punishment level).* It can be viewed as an "everything is possible" result, because it implies that any reasonable payoff can be achieved at equilibrium.

It is difficult to establish who proved the first Folk theorem. It *"has been generally known in the profession for at least 15 or 20 years, but has not been published; its authorship is obscure."* [4].

8.5.1 The Uniform Folk Theorem

Theorem 8.5.1 ("The" Folk theorem) *The set of uniform equilibrium payoffs is the set of feasible and individually rational payoffs:* $E_\infty = E$.

Proof We only have to show that $E \subset E_\infty$. Fix $u \in E$. u is feasible, hence there exists a play $h = (a_1, \ldots, a_t, \ldots)$ such that for each player i, $\frac{1}{T} \sum_{t=1}^{T} g^i(a_t) \to_{T \to \infty} u^i$.

The play h will be called the main path of the strategy, and playing according to h for some player i at stage t means to play the i-component of a_t. For each pair of distinct players (i, j), we fix $x^{i,j}$ in $\Delta(A^j)$ such that $(x^{i,j})_{j \neq i}$ achieves the minimum in the definition of v^i. Fix now a player i in N. We define his strategy σ^i.

σ^i plays at stage 1 according to the main path, and continues to play according to h as long as all the other players do. If at some stage $t \geqslant 1$, for the first time some player j does not follow the main path, then σ^i plays at all the following stages the mixed action $x^{j,i}$ (if for the first time several players leave the main path at the same stage, by convention the smallest player, according to any fixed linear order on N defined in advance, is punished). Since u is in IR, it is easy to see that $\sigma = (\sigma^i)_{i \in N}$ is a uniform equilibrium of G_∞ with payoff u. $\qquad \Box$

8.5.2 The Discounted Folk Theorem

Let us now consider discounted evaluations of payoffs, and come back to the prisoner's dilemma of Example 8.2.3 with discount rate $\lambda \in (0, 1]$. We wonder if $(3, 3) \in E_\lambda$. Playing D^i for the first time will increase once the stage payoff of player i by 1 unit, but then at each stage the payoff of this player will be reduced by 2. Hence with the standard Folk theorem strategies of Theorem 8.5.1, $(3, 3)$ will be in G_λ if $1 \leqslant 2 \sum_{t=1}^{\infty} (1 - \lambda)^t = 2(1 - \lambda)/\lambda$, hence if the players are sufficiently patient in the sense that $\lambda \leqslant 2/3$.

In general we have $E_\lambda \subset E_\infty = E$ for each λ, and the convergence of E_λ to E when patience becomes extreme (λ goes to 0 or δ goes to 1) is worth studying. This convergence (and the convergences in the forthcoming Theorems 8.5.3, 8.5.6, 8.5.8) should be understood with respect to the Hausdorff distance between non-empty compact subsets of $\mathbb{R}^N$, defined by

$$d(A, B) = \max\{\sup_{a \in A} \inf_{b \in B} \|a - b\|, \sup_{b \in B} \inf_{a \in A} \|a - b\|\}.$$

Theorem 8.5.2 (Discounted Folk theorem [198]) *Assume there are two players, or that there exists* $u = (u^i)_{i \in N}$ *in E such that for each player i, we have* $u^i > v^i$. *Then* $E_\lambda \xrightarrow[\lambda \to 0]{} E$.

Proof Since $E_\lambda \subset E$ for each λ, all we have to show is: for all $\varepsilon > 0$, there exists a $\lambda_0 > 0$ such that for all $\lambda \in (0, \lambda_0)$, each point in E is ε-close to a point in E_λ.

(1) Fix $\varepsilon > 0$. There exists a $\bar{\lambda} > 0$ such that for each $\lambda \leqslant \bar{\lambda}$ and each $\hat{u}$ in E, one can find a periodic sequence $h = (a_1, \ldots, a_t, \ldots)$ of points in A satisfying

$$\forall i \in N, \forall T \geqslant 1, \left| \hat{u}^i - \sum_{t=T}^{\infty} \lambda(1 - \lambda)^{t-T} g^i(a_t) \right| \leqslant \varepsilon.$$

If the players follow h by playing $a_1, \ldots, a_t, \ldots$, not only is the discounted payoff $u^\dagger := \sum_{t=1}^{\infty} \lambda(1 - \lambda)^{t-1} g(a_t)$ ε-close to $\hat{u}$, but all continuation discounted payoffs are.

Consider now $\hat{u}$ in E such that $\hat{u}^i \geqslant v^i + 2\varepsilon$ for each i. Given a periodic sequence h as above, consider the strategy profile where all players follow h, and deviations of a player i at some stage are punished by all other players at his minmax level v^i at all subsequent stages. We have for each stage T:

$$\sum_{t=T}^{\infty} \lambda(1 - \lambda)^{t-T} g^i(a_t) \geqslant \hat{u}^i - \varepsilon \geqslant v^i + \varepsilon \geqslant \lambda \|g\|_\infty + (1 - \lambda)v^i,$$

assuming that λ is also small enough such that $\lambda(\|g\|_\infty - v^i) \leqslant \varepsilon$ for each i. Hence at every stage T, the continuation payoff without deviating is at least as large as the continuation payoff in case of deviation (at most $\|g\|_\infty$ at stage T, and v^i at all following stages). No deviation if profitable, and $u^\dagger$ belongs to E_λ.

We have obtained that for each $\varepsilon > 0$, there exists a $\lambda_0 > 0$ such that for all $\lambda \in (0, \lambda_0)$, each point $\hat{u}$ in E such that $\hat{u}^i \geqslant v^i + 2\varepsilon$ for each i is ε-close to a point in E_λ.

(2) Assume that there exists a $u_* = (u^i_*)_{i \in N}$ in E and $\beta > 0$ such that for each player i, we have $u^i_* \geqslant v^i + \beta$.

Since E is convex and bounded, there exists a $K > 0$ such that for ε in $(0, \beta]$ we have

$$\forall u \in E, \exists \hat{u} \in E \text{ s.t. :} \ \|u - \hat{u}\|_\infty \leqslant K\varepsilon \text{ and } \forall i \in N, \hat{u}^i \geqslant v^i + 2\varepsilon.$$

(Proof: let K be such that $2\|u' - u''\|_\infty \leqslant \beta K$ for all u', u'' in E, and define $\hat{u} = (1 - \frac{2\varepsilon}{\beta})u + \frac{2\varepsilon}{\beta}u_*$.)

Fix ε in $(0, \beta]$, and consider any u in E. There exists a $\hat{u} \in E$ such that $\|u - \hat{u}\|_\infty \leqslant K\varepsilon$ and $\hat{u}^i \geqslant v^i + 2\varepsilon$ for each i. By part (1), there exists a λ_0, only depending on ε, such that for all $\lambda \in (0, \lambda_0)$, $\hat{u}$ is ε-close to a point in E_λ. Hence $d(u, E_\lambda) \leqslant (K + 1)\varepsilon$, and $d(E, E_\lambda) \leqslant (K + 1)\varepsilon$. This is true for all $\varepsilon > 0$, hence $E_\lambda \xrightarrow[\lambda \to 0]{} E$.

(3) Assume finally that there are two players, and that we are not in case (2). By convexity of E, we can assume without loss of generality that $\forall u \in E, u^1 = v^1$. The case $E = \{v\}$ is easy, so let us assume that $E \neq \{v\}$. By convexity of E again, the set of feasible payoffs $cog(A)$ is a subset of $H := \{u \in \mathbb{R}^2, u^1 \leqslant v^1\}$, and each point in E is a convex combination of points u in $g(A)$ such that $u^1 = v^1$. One can proceed as in part (1), with periodic sequences $h = (a_1, \ldots, a_t, \ldots)$ of points in A such that $g^1(a_t) = v^1$ for each t, and no need to punish deviations by player 1, and obtain: for each $\varepsilon > 0$, there exists a λ_0 such that for all $\lambda \in (0, \lambda_0)$, each point $u = (v^1, u^2)$ in E such that $u^2 \geqslant v^2 + 2\varepsilon$ is ε-close to a point in E_λ. The result follows. □

The following example (due to Forges, Mertens and Neyman [60]) is a counter-example to the convergence of E_λ to E:

(1, 0, 0)	(0, 1, 0)
(0, 1, 0)	(1, 0, 1)

Player 1 chooses a row, player 2 chooses a column and player 3 chooses nothing here. Essentially, this game is a zero-sum game between players 1 and 2, and in each Nash equilibrium of G_λ each of these players independently randomizes at every stage between his two actions with equal probabilities. Therefore $E_\lambda = \{(1/2, 1/2, 1/4)\}$ for each λ, whereas $(1/2, 1/2, 1/2) \in E$.

8.5.3   The Finitely Repeated Folk Theorem

We now consider finitely repeated games. In the prisoner's dilemma, we have seen that for each T, $E_T = \{(1, 1)\}$. Hence the Pareto-optimal equilibrium payoff $(3, 3)$ cannot be approximated by equilibrium payoffs of finitely repeated games, and there is no convergence of E_T to E.

Theorem 8.5.3 (The finitely repeated Folk theorem [19]) *Assume that for each player i there exists a u in E_1 such that $u^i > v^i$.*
    Then $E_T \xrightarrow[T \to \infty]{} E$.

Proof Since $E_T \subset E$ for each T, all we have to show is: for all $\varepsilon > 0$, there exists a $T_0 > 0$ such that for all $T \geqslant T_0$, each point in E is ε-close to a point in E_T. By assumption there exists a $\beta > 0$ and for each player i in N, a Nash equilibrium payoff $u(i)$ in E_1 such that $u^i(i) \geqslant v^i + \beta$.

The idea is to play a long main path of pure actions, where deviations are punished by perpetual minmaxing, followed by several rounds of "good" Nash equilibria $u(i)$ for each player i, in order to reward the players in case no deviation occurs in the main path.

Fix $\varepsilon > 0$. There exists an $L \geqslant 1$ such that each point u in E is ε-close to a point $\hat{u}$ in E of the form $\hat{u} = \frac{1}{L} \sum_{t=1}^{L} g(a_t)$, with $a_1, \ldots, a_L$ in A. Fix such L, then fix R such that $R\beta \geqslant \|g\|_\infty (L+2) + L \max_i v^i$ and finally fix K large enough such that $\frac{1}{K} \leqslant \varepsilon$ and $\frac{nR}{KL} \leqslant \varepsilon$ (where n is the number of players). Consider a T-stage repeated game, with $T \geqslant KL + nR$. By Euclidean division, T can be written as $T = K'L + nR + L'$, with $L' \in \{0, \ldots, L-1\}$ and $K' \geqslant K$.

Consider now u in E, and let $\hat{u} = \frac{1}{L} \sum_{t=1}^{L} g(a_t)$ be as above. Define the strategy profile σ in the game with $T = K'L + nR + L'$ stages as follows:

In the main path, the players: first play K' blocks of length L, in each of these blocks the sequence $a_1, \ldots, a_L$ is played, then play in a fixed given order, for each player i, R times the Nash equilibrium of the one-shot game with payoff $u(i)$, and finally play for the remaining L' stages an arbitrary fixed Nash equilibrium x of the one-shot game.

If a player deviates from the main path at some stage T, he is punished by the other players at all subsequent stages at his minmax level v^i.

We claim that σ is a Nash equilibrium of the T-stage repeated games. Indeed, suppose player i deviates from the main path at stage $t = kL + l$, with $k \in \{0, K'-1\}$ and $l \in \{1, \ldots, L\}$. The sum of his payoffs, from stage t to stage T, is then at most $\|g\|_\infty + (T-t)v^i$. If he follows the main path, this sum would be at least $-\|g\|_\infty (L-l+1) + (T-(k+1)L)v^i + R\beta$. Since $R\beta \geqslant \|g\|_\infty (L+2) + Lv^i$, we have $-\|g\|_\infty (L-l+1) + (T-(k+1)L)v^i + R\beta \geqslant \|g\|_\infty + (T-t)v^i$ and the deviation is not profitable. Later deviations from the main path are also not profitable since one-shot Nash equilibria are played at the last $nR + L'$ stages. We obtain that σ is a Nash equilibrium, and its payoff $u^\dagger$ belongs to E_T. We have

$$u^\dagger = \frac{K'L}{T} \hat{u} + \frac{R}{T} \sum_{i \in N} u(i) + \frac{L'}{T} g(x),$$

and since $T = K'L + nR + L'$,

$$\|\hat{u} - u^\dagger\|_\infty \leqslant \frac{(nR + L')}{T} (\|\hat{u}\|_\infty + \|g\|_\infty) \leqslant 2\frac{(nR + L)}{T} \|g\|_\infty.$$

Since $T \geqslant KL$, the assumptions on K imply $\|\hat{u} - u^\dagger\|_\infty \leqslant 4\varepsilon \|g\|_\infty$. Since u is ε-close to $\hat{u}$, $\|u - u^\dagger\|_\infty \leqslant \varepsilon(1 + 4\|g\|_\infty)$, and the result follows. $\qquad\square$

8.5.4   Subgame-Perfect Folk Theorems

In the equilibria constructed in the previous proofs, after some player has first deviated from the main path, it might not be profitable for the other players to punish him. One can then define, as in extensive-form games, the notion of a subgame-perfect equilibrium. Given a history h in H and a strategy profile σ, the continuation strategy $\sigma[h]$ is defined as the strategy profile $\tau = (\tau^i)_{i \in N}$, where $\forall i \in N, \forall h' \in H : \tau^i(h') = \sigma^i(hh')$, with the notation that hh' is the concatenation of history h followed by history h'.

8.5.4.1   Subgame-Perfect Uniform Equilibria

Definition 8.5.4 A *subgame-perfect uniform equilibrium* is a strategy profile σ in Σ such that for each history h in H, $\sigma[h]$ is a uniform equilibrium of G_∞. We denote by E'_∞ the set of payoff profiles of these equilibria.

Considering $h = \varnothing$ implies that a subgame-perfect uniform equilibrium is a uniform equilibrium, hence $E'_\infty \subset E_\infty = E$. In 1976, Aumann and Shapley, as well as Rubinstein (see the re-editions from [14, 183]), proved slight variants of the following result: introducing subgame-perfection changes nothing to the Folk theorem for uniform equilibria.

Theorem 8.5.5 (Perfect Folk Theorem)

$$E'_\infty = E_\infty = E.$$

Proof Given a feasible and individually rational payoff, we have to construct a subgame-perfect uniform equilibrium. In comparison with the proof of Theorem 8.5.1, it is sufficient to modify the punishment phase. If at some stage t, for the first time player j leaves the main path, the other players $-j$ will punish player j (again, if for the first time several players leave the main path at the same stage, by convention the smallest player, according to any fixed linear order on N defined in advance, is punished), but now the punishment will not last forever. It will last until some stage $\bar{t}$, and then, whatever has happened during the punishment phase, all players forget everything and come back, as in stage 1, onto the main path. One possibility is to define $\bar{t}$ so that the expected average payoff of player j up to stage $\bar{t}$ will be lower than $v^j + 1/t$. Another possibility is to simply put $\bar{t} = 2t$. $\square$

8.5.4.2   Subgame-Perfect Discounted Equilibria

Similarly, one can define the subgame-perfect equilibria of G_λ as Nash equilibria of every subgame of G_λ, and we denote by E'_λ the set of subgame-perfect equilibrium payoffs. E'_λ is non-empty since it contains the mixed Nash equilibrium payoffs of the stage game G, and compact since the set of subgame-perfect equilibria of G_λ is an intersection of compact sets (one for each subgame).

Theorem 8.5.6 (Perfect discounted Folk theorem [69]) *If E has non-empty interior, then $E'_\lambda \xrightarrow[\lambda \to 0]{} E$.*

An example where E has empty interior and the convergence of E'_λ to E fails is the following two-player game:

(1, 0)	(1, 1)
(0, 0)	(0, 0)

In each subgame-perfect equilibrium of the discounted game, player 1 chooses the Top row at every stage after any history, so $E'_\lambda = \{(1, 1)\}$ for each λ. However, $(1, 0) \in E$.

The proof of Theorem 8.5.6 uses punishments and also "rewards" to give incitations for the players to punish someone who deviated. The details are a bit technical, and we mostly develop here a sketch of Mertens, Sorin and Zamir [137, Exercise 8, Sect. IV.4] (see also [124, Sect. 3.8], or [68]). We start with a lemma, where $B_\infty(d, r)$ denotes the open ball of center d and radius r for $\|\cdot\|_\infty$. We give a complete proof of the lemma in Exercise 11 at the end of this chapter.

Lemma 8.5.7 *Given $\varepsilon > 0$ there exists a $\lambda_0 > 0$ such that for every $\lambda \leqslant \lambda_0$, for every player i in N and feasible payoff $d = (d^j)_{j \in N}$ such that $B_\infty(d, \varepsilon) \subset \text{cog}(A)$, there exists a path $(a_t)_{t \geqslant 1}$ of elements in A such that*

$$d = \sum_{t=1}^{\infty} \lambda(1 - \lambda)^{t-1} g(a_t)$$

(d is the λ-discounted payoff of the path), and for each $T \geqslant 1$,

$$\left\| \sum_{t=T}^{\infty} \lambda(1 - \lambda)^{t-T} g(a_t) - d \right\|_\infty \leqslant \varepsilon, \text{ and } \sum_{t=T}^{\infty} \lambda(1 - \lambda)^{t-T} g^i(a_t) \geqslant d^i$$

(all continuation payoffs are close to d, and for player i they are at least d^i).

Proof of Theorem 8.5.6 We will actually prove that for each $\varepsilon > 0$, there exists a λ_0 such that if $0 < \lambda \leqslant \lambda_0$, every payoff d such that $B_\infty(d, \varepsilon) \subset E$ is a subgame-perfect equilibrium payoff of G_λ. This will be enough to prove the theorem since E is convex with non-empty interior.

Fix ε, and a positive integer L such that $L\varepsilon \geqslant 3\|g\|_\infty$. Consider $d = (d^i)_{i \in N}$ such that $B_\infty(d, 5\varepsilon) \subset E$. By Lemma 8.5.7, there exists a path $(a_t)_{t \geqslant 1}$ of elements in A such that $d = \sum_{t=1}^{\infty} \lambda(1 - \lambda)^{t-1} g(a_t)$, and all continuation payoffs are ε-close to d. For any history $l = (l_t)_{t=1,\dots,L}$ in A^L and player i in N, we define $d(l, i)$ in $\mathbb{R}^N$ by: $d^i(l, i) = d^i - 4\varepsilon$ $(\geqslant v^i + \varepsilon)$, and for each player $j \neq i$, $d^j(l, i)$ is defined by the equation

$$d^j = \sum_{t=1}^{L} \lambda(1-\lambda)^{t-1}g^j(l_t) + (1-\lambda)^L d^j(l,i). \tag{8.2}$$

For λ small enough (depending only on ε and L), $|d^j - d^j(l,i)| \leqslant \varepsilon$ for $j \neq i$, and $\|d - d(l,i)\|_\infty \leqslant 4\varepsilon$, so we can apply Lemma 8.5.7 again and obtain a path $h(l,i) = (a_t(l,i))_{t \geqslant 1}$ satisfying the conclusions of the lemma for $d(l,i)$.

We will construct an equilibrium starting with the path $(a_t)_{t \geqslant 1}$. If a player i deviates from the current path, he will be punished for L stages, then the path $h(l,i)$ will be played. When player i is punished, randomizations by the other players may be required, so it is difficult to check for deviations of these players. However Eq. (8.2) will ensure that they have the same continuation payoffs, so are indifferent between all histories during the punishment phase. Formally, we define the strategy profile σ by:

Initialization: $h = (a_t)_{t \geqslant 1}$. Start with Phase 1.

Phase 1: All players follow the path of pure actions h. As soon as a single player i deviates from h, go to phase $2i$. If several players deviate from h at the same stage, ignore the deviations.

Phase 2i: During L independent stages, all players $j \neq i$ punish player i to his independent minmax, and player i plays a best response to the punishment. This punishment phase defines a history l in A^L of action profiles played. Go back to phase 1, with $h = h(l,i)$.

We show that σ is a subgame-perfect equilibrium, provided that λ is small enough (independently of d). By the *one-shot deviation principle* (see question 1) of Exercise 10 at the end of this chapter), it is enough to consider deviations at a single stage.

(1) During the first phase 1, when $(a_t)_{t \geqslant 1}$ is played:

If a player i conforms to the strategy, his continuation payoff is at least $d^i - \varepsilon$. If player i deviates, his continuation payoff is at most

$$\lambda\|g\|_\infty + (1-\lambda)((1-(1-\lambda)^L)v^i + (1-\lambda)^L(d^i - 4\varepsilon)).$$

Since $v^i \leqslant d^i - 5\varepsilon$, the deviation is not profitable when λ is small enough (such that $\lambda(2\|g\|_\infty + 4\varepsilon) \leqslant 3\varepsilon$).

(2) During phase 1, with $h = h(l,i)$ for some i:

If a player $j \neq i$ conforms to the strategy, his continuation payoff is at least $d^j(l,i) - \varepsilon \geqslant d^j - 2\varepsilon$. If j deviates, his continuation payoff is at most

$$\lambda\|g\|_\infty + (1-\lambda)((1-(1-\lambda)^L)v^j + (1-\lambda)^L(d^j - 4\varepsilon)).$$

Again, $v^j \leqslant d^j - 5\varepsilon$ ensures that the deviation is not profitable when λ is small enough.

If player i conforms to the strategy, his continuation payoff is at least $d^i(l,i)$ by the property of $h(l,i)$ coming from the last inequality of Lemma 8.5.7. If player i deviates, his payoff is at most

$$\lambda \|g\|_\infty + (1 - \lambda)((1 - (1 - \lambda)^L)v^i + (1 - \lambda)^L(d^i(l, i)).$$

Since $v^i \leqslant d^i(l, i) - \varepsilon$, the deviation is not profitable if

$$2\lambda \|g\|_\infty \leqslant \varepsilon(1 - \lambda)(1 - (1 - \lambda)^L).$$

The right-hand side is equivalent to $\varepsilon\lambda L$ when λ goes to 0, hence the definition of L ensures that this deviation is not profitable for λ small.

(3) During phase $2i$:

Each player $j \neq i$ has, at every stage, a continuation payoff which is independent of the remaining history to be played during this phase, so no deviation can be profitable. Player i is punished at every stage, then has a continuation payoff of $d^i(l, i) = d^i - 4\varepsilon$ which is independent of the history l eventually played in this phase, so he cannot do better than best replying at every stage to the punishment.

No deviation is thus profitable, and σ is a subgame perfect equilibrium of the λ-discounted game. $\qquad \square$

8.5.4.3 Subgame-Perfect Finitely Repeated Equilibria

Finally, one defines the subgame-perfect equilibria of G_T as usual, i.e. as strategy profiles σ such that $\forall t \in \{0, \ldots, T - 1\}, \forall h \in H_t, \sigma[h]$ is a Nash equilibrium of the game starting at h, i.e. of the game G_{T-t}. We denote by E'_T the (non-empty and compact) set of subgame-perfect equilibria of G_T.

Theorem 8.5.8 (Perfect finitely repeated Folk theorem [81]) *Assume that for each player i, there exists x and y in E_1 such that $x^i > y^i$, and that E has a non-empty interior. Then $E'_T \xrightarrow[T \to \infty]{} E$.*

Gossner's proof includes main paths, punishments and rewards. It starts with a Folk theorem for finitely repeated games *with terminal payoffs*, using statistical test functions to induce effective punishment during the punishment phases. It is omitted here.

We conclude this section with a standard example of a subgame-perfect discounted equilibrium from economics.

Example 8.5.9 (*Repeated Bertrand competition*) Consider an oligopoly with n identical firms, producing the same good with constant marginal cost $c > 0$. Each of the firms has to fix a selling price, and the consumers will buy the good from the firm with the lowest price (in case of several lowest prices, the firms with that price will equally share the market). We denote by $D(p)$ the number of consumers willing to buy a unit of good at price p and assume that the demand is always fulfilled. Each firm wants to maximize its profit, which is $\pi(p) = D(p)(p - c)$ if the firm is alone with the lowest price p, and $\pi(p) = 0$ if the firm sells nothing. Assume that π has a maximum at some price $\hat{p} > c$.

If the game is played once, there is a unique equilibrium price which is the marginal cost c, and the profits are zero. Let us introduce dynamics (firms can adapt their price depending on the competition) and consider the repeated game with discount factor λ. Define a strategy profile where everyone plays $\hat{p}$ as long as everybody does so, and if there is a deviation everyone plays c forever. The payoff of a firm if all firms follow this strategy is $\pi(\hat{p})/n$, and the payoff of a firm deviating from this strategy by playing a price p at some stage will be, from that stage on, at most $\lambda\pi(p) + (1 - \lambda)0 = \lambda\pi(p)$. Hence if the players are sufficiently patient in the sense that $\lambda \leqslant 1/n$, we have a subgame-perfect equilibrium where the realized price is the collusion (or monopoly) price $\hat{p}$.

8.6 Extensions: Stochastic Games, Incomplete Information, Signals

In the following sections, we present a few examples of general repeated games (or *Markovian Dynamic Games*). In each of them, we assume that players always remember their own past actions and signals received, so that they are assumed to have perfect recall. In every case, a strategy σ^i for player i is a sequence $(\sigma^i_t)_{t \geqslant 1}$, where σ^i_t gives the mixed action played by player i at stage $t + 1$ as a function of the past history observed by this player at previous stages. A strategy profile $(\sigma^i)_i$ naturally induces a probability distribution over plays (infinite joint histories), and one can define as in Sect. 8.3.3 the T-stage game, the λ-discounted game and the uniform game. Since we will always consider games with finite sets of players, actions, states and signals, the existence of Nash equilibria (value and optimal strategies in the zero-sum case) is ensured in the T-stage game and in the λ-discounted game.

8.6.1 A Repeated Game with Signals

In a repeated game with signals (also called with imperfect observation or with imperfect monitoring), players do not perfectly observe after each stage the action profile that has been played, but receive private signals depending on this action profile. In the following example there are two players, and the sets of signals are given by $U^1 = \{u, v, w\}$ and $U^2 = \{*\}$. This means that after each stage player 1 will receive a signal in $\{u, v, w\}$ whereas player 2 will always receive the signal $*$, which is equivalent to receiving no signal at all: the observation of player 2 is said to be *trivial*. Payoffs in the stage game and signals for player 1 are given by:

	L	R
T	$(0, 0), u$	$(4, 5), v$
B	$(1, 0), w$	$(5, 0), w$

For example, if at some stage player 1 plays T and player 2 plays R then the stage payoff is $(4, 5)$, and before playing the next stage player 1 receives the signal v (hence he can deduce that player 2 has played R and compute the payoffs) whereas player 2 observes nothing (and in particular is not able to deduce his payoff). Here $(4, 5)$ is feasible and individually rational. However, since player 2 has trivial observation, it is dominant for player 1 in the repeated game to play B at every stage. At equilibrium it is not possible to play (T, R) a significant number of times, otherwise player 1 could profitably deviate by playing B without being punished afterwards. Formally, one can easily show here that $E_\infty = \text{co}\{(1, 0), (5, 0)\}$, and thus is a strict subset of the set of feasible and individually rational payoffs E.

Computing E_∞ or $\lim_{\lambda \to 0} E_\lambda$ for repeated games with signals is not known in general, even for two players. There is a wide literature on this topic. In particular, Lehrer [119, 120], introduces the notions of indistinguishable and more informative mixed actions, Renault and Tomala [170] give a general characterization of uniform communication equilibrium payoffs in repeated games with signals (generalizing the notion of individually rational payoffs to jointly rational payoffs, see Exercise 9), and Gossner and Tomala [82] use entropy methods to compute the minmax levels in some repeated games with signals. In repeated games with imperfect public observation, Abreu, Pearce and Stacchetti [1] provide, for a given discount factor, a characterization of the perfect public equilibrium payoff set as the largest bounded fixed point of an operator (see Exercise 10), and Fudenberg, Levine and Maskin [68] provide conditions of full rank and pairwise-identifiability implying the Folk theorem, in the sense of convergence of E_λ to the set of feasible and individually rational payoffs.

8.6.2 A Stochastic Game: The Big Match

Stochastic games were introduced by Shapley in [188]. In these dynamic interactions there are several possible states, and to each state is associated a strategic game. The state may evolve from stage to stage, and at every stage the game corresponding to the current state is played. The action profile and the current state determine the current payoffs and also the transition to the next state.

The following example is a two-player zero-sum game called the "Big Match", introduced by Gillette [77].

$$
\begin{array}{c|c|c|}
 & L & R \\
\hline
T & 1* & 0* \\
\hline
B & 0 & 0 \\
\hline
\end{array}
$$

The players start at stage 1 by playing the above matrix. They continue as long as player 1 plays B (and both players observe after each stage the action played by their opponent). However, if at some stage player 1 plays T, then the game stops and either at that stage player 2 has played L and player 1 has a payoff of 1 at all future stages, or player 2 has played R at that stage and player 1 has a payoff of 0 at all future stages. Formally, there are three states here: the initial state, the state where player 1 has payoff 0 whatever happens and the state where player 1 has payoff 1 whatever happens. The last two states are said to be *absorbing*, and once an absorbing state is reached the play stays there forever (absorbing states are represented by a $*$ in the matrix).

Denote by v_T the value of the stochastic game with T stages (and average payoff $\mathbb{E}_\sigma(\frac{1}{T}\sum_{t=1}^{T} g_t)$) and by v_λ the value of the discounted stochastic game. Shapley [188] shows that dynamic programming naturally extends and leads to the Shapley formulas: $\forall T \geqslant 1, \forall \lambda \in (0, 1]$,

$$(T+1)v_{T+1} = \mathrm{val}\begin{pmatrix} T+1 & 0 \\ Tv_T & 1+Tv_T \end{pmatrix},$$

$$v_\lambda = \mathrm{val}\begin{pmatrix} 1 & 0 \\ (1-\lambda)v_\lambda & 1+\lambda+(1-\lambda)v_\lambda \end{pmatrix},$$

and one easily gets $v_T = v_\lambda = 1/2$ for each T and λ.

Suppose now that the players do not precisely know the large number of stages to be played. Player 2 can play at every stage the mixed action $1/2\, L + 1/2\, R$, and by doing so he guarantees an expected payoff of $1/2$ at every stage. Hence $\exists \sigma^2, \forall T, \forall \sigma^1, \gamma_T^1(\sigma^1, \sigma^2) \leqslant 1/2$. Here, it is more difficult and fascinating to figure out good long-term strategies for player 1.

Theorem 8.6.1 ([28]) *The Big Match has a uniform value which is 1/2, i.e.:*

Player 1 uniformly guarantees 1/2, i.e. $\forall \varepsilon > 0, \exists \sigma^1, \exists T_0, \forall T \geqslant T_0, \forall \sigma^2, \gamma_T^1(\sigma^1, \sigma^2) \geqslant 1/2 - \varepsilon$; *and*

Player 2 uniformly guarantees 1/2, i.e. $\forall \varepsilon > 0, \exists \sigma^2, \exists T_0, \forall T \geqslant T_0, \forall \sigma^1, \gamma_T^1(\sigma^1, \sigma^2) \leqslant 1/2 + \varepsilon$.

If the number of repetitions is large enough, known or possibly unknown to the players, the fair "price per period" for this game is $1/2$.

Proof (the redaction here follows [168]) We need to show that player 1 uniformly guarantees $1/2$. First define the following random variables, for all $t \geqslant 1$: g_t is the payoff of player 1 at stage t, $i_t \in \{T, B\}$ is the action played by player 1 at stage t, $j_t \in \{L, R\}$ is the action played by player 2 at stage t, $L_t = \sum_{s=1}^{t-1} \mathbf{1}_{j_s=L}$ is the number of stages in $1, \ldots, t-1$ where player 2 has played L, $R_t = \sum_{s=1}^{t-1} \mathbf{1}_{j_s=R} = t-1-L_t$ is the number of stages in $1, \ldots, t-1$ where player 2 has played R, and $m_t = R_t - L_t \in \{-(t-1), \ldots, 0, \ldots, t-1\}$. $R_1 = L_1 = m_1 = 0$.

Given a positive integer parameter M, let us define the following strategy σ_M^1 of player 1: at any stage t, σ_M^1 plays T with probability $\frac{1}{(m_t+M+1)^2}$, and B with the remaining probability. Some intuition for σ_M^1 can be given. Assume we are still in the non-absorbing state at stage t. If player 2 has played R often at past stages, player 1 is doing well and has received good payoffs, m_t is large and σ_M^1 plays the risky action T with small probability. On the other hand if player 2 is playing L often, player 1 has received low payoffs but player 2 is taking high risks; m_t is small and σ_M^1 plays the risky action T with high probability.

Notice that σ_M^1 is well defined. If $m_t = -M$ then σ_M^1 plays T with probability 1 at stage t and then the game is over. So the event $m_t \leqslant -M - 1$ has probability 0 as long as the play is in the non-absorbing state. At any stage t in the non-absorbing state, we have $-M \leqslant m_t \leqslant t - 1$, and σ_M^1 plays T with a probability in the interval $[\frac{1}{(M+t)^2}, 1]$.

We will show that σ_M^1 uniformly guarantees $\frac{M}{2(M+1)}$, which is close to 1/2 for M large. More precisely we will prove that

$$\forall T \geqslant 1, \forall M \geqslant 0, \forall \sigma^2, \quad \mathbb{E}_{\sigma_M^1, \sigma^2}\left(\frac{1}{T}\sum_{t=1}^T g_t\right) \geqslant \frac{M}{2(M+1)} - \frac{M}{2T}. \tag{8.3}$$

We now prove (8.3). Notice that we can restrict attention to strategies of player 2 which are pure, and (because there is a unique relevant history of moves of player 1) independent of the history. That is, we can assume w.l.o.g. that player 2 plays a fixed deterministic sequence $y = (j_1, \ldots, j_t, \ldots) \in \{L, R\}^\infty$.

T being fixed until the end of the proof, we define the random variable t^* as the time of absorption:

$$t^* = \inf\{s \in \{1, \ldots, T\}, i_s = T\}, \text{ with the convention}$$
$$t^* = T + 1 \text{ if } \forall s \in \{1, \ldots, T\}, i_s = B.$$

Recall that $R_t = m_t + L_t = t - 1 - L_t$, so that $R_t = \frac{1}{2}(m_t + t - 1)$. For $t \leqslant t^*$, we have $m_t \geqslant -M$, so

$$R_{t^*} \geqslant \frac{1}{2}(t^* - M - 1).$$

Also define X_t as the following fictitious payoff of player 1: $X_t = 1/2$ if $t \leqslant t^* - 1$, $X_t = 1$ if $t \geqslant t^*$ and $j_{t^*} = L$, and $X_t = 0$ if $t \geqslant t^*$ and $j_{t^*} = R$. X_t is the random variable of the limit value of the current state. A simple computation shows

$$\mathbb{E}_{\sigma_M^1, y}\left(\frac{1}{T}\sum_{t=1}^T g_t\right) = \mathbb{E}_{\sigma_M^1, y}\frac{1}{T}(R_{t^*} + (T - t^* + 1)\mathbf{1}_{j_{t^*}=L})$$

$$\geqslant \mathbb{E}_{\sigma_M^1, y}\frac{1}{T}\left(\frac{1}{2}(t^* - M - 1) + (T - t^* + 1)\mathbf{1}_{j_{t^*}=L}\right)$$

$$\geqslant -\frac{M}{2T} + \mathbb{E}_{\sigma_M^1, y}\frac{1}{T}\left(\frac{1}{2}(t^* - 1) + (T - t^* + 1)\mathbf{1}_{j_{t^*}=L}\right)$$

$$\geqslant -\frac{M}{2T} + \mathbb{E}_{\sigma_M^1, y}\left(\frac{1}{T}\sum_{t=1}^T X_t\right).$$

To prove (8.3), it is thus enough to show the following lemma.

Lemma 8.6.2 *For all t in $\{1, \ldots, T\}$, y in $\{L, R\}^\infty$ and $M \geqslant 1$,*

$$\mathbb{E}_{\sigma_M^1, y}(X_t) \geqslant \frac{M}{2(M+1)}.$$

Proof of the lemma. The proof is by induction on t. For $t = 1$,

$$\mathbb{E}_{\sigma_M^1, y}(X_1) = \frac{1}{2}\left(1 - \frac{1}{(M+1)^2}\right) + \frac{1}{(M+1)^2}\mathbb{1}_{j_1=L}$$

$$\geqslant \frac{1}{2}\left(1 - \frac{1}{(M+1)^2}\right) \geqslant \frac{M}{2(M+1)}.$$

Assume the lemma is true for $t \in \{1, \ldots, T-1\}$. Consider $y = (j_1, \ldots)$ in $\{L, R\}^\infty$, and write $y = (j_1, y_+)$ with $y_+ = (j_2, j_3, \ldots) \in \{L, R\}^\infty$. If $j_1 = L$,

$$\mathbb{E}_{\sigma_M^1, y}(X_{t+1}) = \frac{1}{(M+1)^2}1 + (1 - \frac{1}{(M+1)^2})\mathbb{E}_{\sigma_{M-1}, y_+}(X_t).$$

By the induction hypothesis, $\mathbb{E}_{\sigma_{M-1}, y_+}(X_t) \geqslant \frac{M-1}{2M}$, so

$$\mathbb{E}_{\sigma_M^1, y}(X_{t+1}) \geqslant \frac{M}{2(M+1)}.$$

Otherwise $j_1 = R$, and

$$\mathbb{E}_{\sigma_M^1, y}(X_{t+1}) = \left(1 - \frac{1}{(M+1)^2}\right)\mathbb{E}_{\sigma_{M+1}, y_+}(X_t)$$

$$\geqslant \left(1 - \frac{1}{(M+1)^2}\right)\frac{M+1}{2(M+2)} = \frac{M}{2(M+1)}.$$

Lemma 8.6.2 is proved.   □

This concludes the proof of Theorem 8.6.1.   □

The existence of a uniform value for zero-sum stochastic games (with finite states and actions) has been proved in Mertens and Neyman [136]. The generalization to the existence of equilibrium payoffs in non-zero-sum stochastic games was completed by Vieille [218, 219] for two players, and the problem is still open for more than two players (see Exercise 12, and Solan-Vieille [196], on *quitting games*). There is a wide literature on stochastic games, see the surveys by Mertens [135], Vieille [220], Solan [195] and Laraki–Sorin [118] and the books by Mertens et al. [137], Sorin [203], and Neyman and Sorin [154]. In particular (and among others), Sorin analyses a non-zero sum version of the Big Match [199], Renault and Ziliotto [172] provide an example where the equilibrium payoffs set does not converge when the discount rate goes to zero, Flesch et al. [55] exhibit a three-player quitting game without stationary ε-equilibria and Solan [194] shows the existence of uniform equilibrium payoffs in three-player stochastic games which are absorbing (there is a single state which is non-absorbing, i.e. which can possibly be left by the players). In the zero-sum case, one can ask whether the results can be extended to compact action sets and continuous payoffs and transitions: the answer is positive with some algebraic definability conditions [29], but in the general compact-continuous case $\lim_T v_T$ and $\lim_\lambda v_\lambda$ may even fail to exist [221].

We can mention the existence of several links and common tools between the asymptotic behavior in repeated games (as the discount factor λ goes to zero or the horizon T goes to infinity) and the convergence of time discretizations of some continuous time games (as the mesh goes to zero), see for instance Laraki and Sorin [118]. Finally, observe that there are other ways of evaluating the stream of payoffs in repeated games, see Maitra and Sudderth [125], Martin [126, 127] and Neyman and Sorin [154].

8.6.3   Repeated Games with Incomplete Information: The Cav u Theorem

In a repeated game with incomplete information, there are several states as well, and to each state is associated a possible stage game. One of the states is selected *once and for all* at the beginning of the game, and at each stage the players will play the game corresponding to this state. The state is fixed and does not evolve from stage to stage. The difficulty comes from the fact that the players imperfectly know it: each player initially receives a signal depending on the selected state, and consequently may have incomplete information on the state and on the knowledge of the other players on the state.

In the following examples we have two players, zero-sum, and we assume that there is *lack of information on one side*: initially the state $k \in \{a, b\}$ is selected according to the distribution $p = (1/2, 1/2)$, both players 1 and 2 know p and the payoff matrices G^a and G^b, but only player 1 learns k. Then the game G^k is repeated, and after each stage the actions played are publicly observed. We denote by v_T the value of the T-stage game with average payoffs and will discuss, without proofs, the value of $\lim_{T \to \infty} v_T$. By Theorem 8.6.6 below, this limit will always exist and coincide with the limit $\lim_{\lambda \to 0} v_\lambda$ of the values of the discounted games.

Example 8.6.3 $G^a = \begin{pmatrix} 0 & 0 \\ 0 & -1 \end{pmatrix}$ and $G^b = \begin{pmatrix} -1 & 0 \\ 0 & 0 \end{pmatrix}$.

Easy. Player 1 can play T (Top) if the state is a and B (Bottom) if the state is b. Hence $v_T = 0$ for each T, and $\lim_{T \to \infty} v_T = 0$.

Example 8.6.4 $G^a = \begin{pmatrix} 1 & 0 \\ 0 & 0 \end{pmatrix}$ and $G^b = \begin{pmatrix} 0 & 0 \\ 0 & 1 \end{pmatrix}$.

A "naive" strategy of player 1 would be to play at stage 1 the action T if the state is a, and the action B if the state is b. This strategy is called completely revealing, because when player 1 uses it, player 2 can deduce the selected state from the action of player 1. It is optimal here in the 1-stage game, but very bad when the number of repetitions is large. On the contrary, player 1 can play without using his private information on the state, i.e. use a *non-revealing* strategy: he can consider the average matrix $\frac{1}{2}G^a + \frac{1}{2}G^b = \begin{pmatrix} 1/2 & 0 \\ 0 & 1/2 \end{pmatrix}$, and play at each stage the corresponding optimal strategy $1/2\,T + 1/2\,B$. The value of this matrix being $1/4$, we have: $v_T \geqslant \frac{1}{4}$ for each T.

One can also show here that playing non-revealing, i.e. not using his private information, is the best that player 1 can do in the long run: $v_T \xrightarrow[T \to \infty]{} 1/4$.

Example 8.6.5 $G^a = \begin{pmatrix} 4 & 0 & 2 \\ 4 & 0 & -2 \end{pmatrix}$ and $G^b = \begin{pmatrix} 0 & 4 & -2 \\ 0 & 4 & 2 \end{pmatrix}$.

Playing a completely revealing strategy guarantees nothing positive for player 1, because player 2 can finally play M(iddle) if the state is a, and L(eft) if the state is b. Playing a non-revealing strategy leads to considering the average game $\frac{1}{2}G^a + \frac{1}{2}G^b$

$= \begin{pmatrix} 2 & 2 & 0 \\ 2 & 2 & 0 \end{pmatrix}$, and guarantees nothing positive either.

Here, it is interesting for player 1 to play the following strategy σ^1: initially he selects once and for all an element s in $\{T, B\}$ as follows: if $k = a$, choose $s = T$ with probability $3/4$, and $s = B$ with probability $1/4$; and if $k = b$, choose $s = T$ with probability $1/4$, and $s = B$ with probability $3/4$. Then player 1 plays s at each stage, independently of the moves of player 2.

The conditional probabilities satisfy: $\mathbb{P}(k = a|s = T) = 3/4$, and $\mathbb{P}(k = a|s = B) = 1/4$. Hence at the end of the first stage, player 2 having observed the first move of player 1 will learn something about the state of Nature: his belief will move from $\frac{1}{2}a + \frac{1}{2}b$ to $\frac{3}{4}a + \frac{1}{4}b$ or to $\frac{1}{4}a + \frac{3}{4}b$. But still he does not entirely know the state: there is *partial revelation of information*. The games $\frac{3}{4}G^a + \frac{1}{4}G^b$ and $\frac{1}{4}G^a + \frac{3}{4}G^b$ both have value 1, and respective optimal strategies for player 1 are given by T and by B, they correspond to the actions played by player 1. Hence playing σ^1 guarantees 1 to player 1: $\forall T, \forall \sigma^2, \gamma_T^1(\sigma^1, \sigma^2) \geqslant 1$, and one can prove that player 1 cannot do better here in the long run: $v_T \xrightarrow[T \to \infty]{} 1$.

8.6.3.1 General Case of Incomplete Information on One Side

The next result is due to Aumann and Maschler in the sixties (1966, with a re-edition in [13]). It is valid for any finite set of states K and any collection of payoff matrices $(G^k)_{k \in K}$ of the same size.

Theorem 8.6.6 *In a zero-sum repeated game with lack of information on one side where the initial probability is p and the payoff matrices are given by $(G^k)_{k \in K}$, one has*

$$\lim_{T \to \infty} v_T(p) = \lim_{\lambda \to 0} v_\lambda(p) = \operatorname{cav} u(p),$$

where $u : \Delta(K) \to \mathbb{R}$ is such that, for each probability $q = (q^k)_{k \in K}$, $u(q)$ is the value of the matrix $\sum_{k \in K} q^k G^k$, and $\operatorname{cav} u$ is the smallest concave function above u.

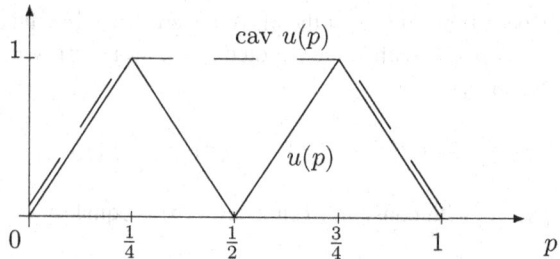

This picture corresponds to Example 8.6.5: cav $u(1/2) = \frac{1}{2} u(1/4) + \frac{1}{2} u(3/4) = 1$. In Example 8.6.4, $u(p) = p(1 - p)$ for each p in $[0, 1]$, so u is concave and cav $u = u$.

Proof The presentation here essentially follows [165]. Denote by M the constant $\max_{k,i,j} |G^k(i, j)|$.

(1) Consider the $T + T'$-stage game with average payoffs. Player 2 can first play an optimal strategy in the T-stage game, then forget everything and play an optimal strategy in the T'-stage game. This implies that: $(T + T')v_{T+T'}(p) \leqslant v_T(p) + v_{T'}(p)$. Since $(v_T(p))_T$ is bounded, it follows that:

$$v_T(p) \xrightarrow[T \to \infty]{} \inf_T v_T(p). \tag{8.4}$$

(2) *Splitting*. The initial probability $p = (p^k)_{k \in K}$ represents the initial belief, or the a priori, of player 2 on the selected state of Nature. Assume that player 1 chooses his first action, or more generally an abstract message or signal s from a finite set S, according to a probability distribution depending on the state, i.e. according to a transition probability $x = (x^k)_{k \in K} \in \Delta(S)^K$. For each signal s, the probability that s is chosen is denoted $\lambda(x, s) = \sum_k p^k x^k(s)$, and given s such that $\lambda(x, s) > 0$ the conditional probability on K, or a posteriori of player 2, is $\hat{p}(x, s) = \left(\frac{p^k x^k(s)}{\lambda(x,s)} \right)_{k \in K}$. We clearly have

$$p = \sum_{s \in S} \lambda(x, s) \hat{p}(x, s), \tag{8.5}$$

so the a priori p lies in the convex hull of the a posteriori. The following lemma expresses a converse: player 1 is able to induce any family of a posteriori containing p in its convex hull. The proof is simple: just put $x^k(s) = \frac{\lambda_s p_s^k}{p^k}$ if $p^k > 0$.

Splitting lemma. *Assume that p is a convex combination $p = \sum_{s \in S} \lambda_s p_s$ with positive coefficients. Then there exists a transition probability $x \in \Delta(S)^K$ such that $\forall s \in S$, $\lambda_s = \lambda(x, s)$ and $p_s = \hat{p}(x, s)$.*

As a consequence, observe that in the T-stage game with a priori p, player 1 can first choose s according to x^k, then play an optimal strategy in the T-stage game with initial probability p_s. This implies that v_T is a concave function of p.

Imagine now that in the T-stage game with a priori p, player 1 decides to reveal no information on the selected state, and plays independently of it. Since payoffs are defined via expectations, it is as if the players were repeating the average matrix game $G(p) = \sum_{k \in K} p^k G^k$, with value denoted $u(p)$. So $v_T(p) \geq u(p)$ for each T, and since v_T is concave,

$$\forall p \in \Delta(K), \forall T \geq 1, \quad v_T(p) \geq \text{cav } u(p). \tag{8.6}$$

Remark u is Lipschitz with constant M and cav $u(p)$ is equal to

$$\max \left\{ \sum_{s \in S} \lambda_s u(p_s), S \text{ finite}, \forall s \ \lambda_s \geq 0, p_s \in \Delta(K), \sum_{s \in S} \lambda_s = 1, \sum_{s \in S} \lambda_s p_s = p \right\}.$$

(3) *Martingale of a posteriori.* Equation (8.5) not only tells that the a posteriori contains p in their convex hull, but also that the expectation of the a posteriori is the a priori. Here we are in a dynamic context, and for every strategy profile σ one can define the process $(p_t(\sigma))_{t \geq 0}$ of the a posteriori of player 2. We have $p_0 = p$, and $p_t(\sigma)$ is the random variable of player 2's belief on the state after the first t stages. More precisely, a strategy profile σ, together with the initial probability p, naturally induces a probability $\mathbb{P}_{p,\sigma}$ over the set of plays $K \times (I \times J)^\infty$, and we define for any $t \geq 0$, $h_t = (i_1, j_1, \ldots, i_t, j_t) \in (I \times J)^t$ and k in K:

$$p_t^k(\sigma, h_t) = \mathbb{P}_{p,\sigma}(k|h_t) = \frac{p^k \mathbb{P}_{k,\sigma}(h_t)}{\mathbb{P}_{p,\sigma}(h_t)}.$$

$p_t(\sigma, h_t) = (p_t^k(\sigma, h_t))_{k \in K} \in \Delta(K)$ (arbitrarily defined if $\mathbb{P}_{p,\sigma}(h_t) = 0$) is the conditional probability on the state of Nature given that σ is played and h_t has occurred in the first t stages. It is easy to see that as soon as $\mathbb{P}_{p,\sigma}(h_t) > 0$, $p_t(\sigma, h_t)$ does not depend on player 2's strategy σ^2, nor on player 2's last action j_t. It is fundamental to observe that:

Martingale of a posteriori: $(p_t(\sigma))_{t \geq 0}$ is a $\mathbb{P}_{p,\sigma}$-martingale with values in $\Delta(K)$.

This is indeed a general property of Bayesian learning of a fixed unknown parameter: *the expectation of what I will know tomorrow is what I know today*. This martingale is controlled by the informed player, and the splitting lemma shows that this player can induce any martingale issued from the a priori p. Notice that, to be able to compute the realizations of the martingale, player 2 needs to know the strategy σ^1 used by player 1.

We now conclude the proof, with the following idea. The martingale $(p_t(\sigma))_{t \geq 0}$ is bounded, hence will converge almost surely, and there is a bound on its L^1 variation (see Lemma 8.6.7 below). This means that after a certain stage the martingale will essentially remain constant, so approximately player 1 will play in a non-revealing way, and so will not be able to have a stage payoff greater than $u(q)$, where q is a "limit a posteriori". Since the expectation of the a posteriori is the a priori p, player 1 cannot guarantee more than

$$\max \left\{ \sum_{s \in S} \lambda_s u(p_s), \ S \text{ finite}, \forall s \in S \ \lambda_s \geq 0, \ p_s \in \Delta(K), \right.$$
$$\left. \sum_{s \in S} \lambda_s = 1, \sum_{s \in S} \lambda_s p_s = p \right\},$$

that is, more than cav $u(p)$. Let us now proceed to the formal proof.

Fix a strategy σ^1 of player 1, and define the strategy σ^2 of player 2 as follows: play at each stage an optimal strategy in the matrix game $G(p_t)$, where p_t is the current a posteriori in $\Delta(K)$. Assume that $\sigma = (\sigma^1, \sigma^2)$ is played in the repeated game with initial probability p. To simplify the notation, we write $\mathbb{P}$ for $\mathbb{P}_{p,\sigma}$, $p_t(h_t)$ for $p_t(\sigma, h_t)$, etc. We use everywhere $\| \cdot \| = \| \cdot \|_1$. To avoid confusion between variables and random variables in the following computations, we will use tildes to denote random variables, e.g. $\tilde{k}$ will denote the random variable of the selected state.

Lemma 8.6.7

$$\forall T \geqslant 1, \quad \frac{1}{T} \sum_{t=0}^{T-1} \mathbb{E}(\| p_{t+1} - p_t \|) \leqslant \frac{\sum_{k \in K} \sqrt{p^k(1 - p^k)}}{\sqrt{T}}.$$

Proof This is a property of martingales with values in $\Delta(K)$ and expectation p. We have for each state k and $t \geqslant 0$:

$$\mathbb{E}\left((p_{t+1}^k - p_t^k)^2\right) = \mathbb{E}(\mathbb{E}((p_{t+1}^k - p_t^k)^2 | \mathcal{H}_t)),$$

where $\mathcal{H}_t$ is the σ-algebra on plays generated by the first t action profiles. So

$$\mathbb{E}\left((p_{t+1}^k - p_t^k)^2\right) = \mathbb{E}(\mathbb{E}((p_{t+1}^k)^2 + (p_t^k)^2 - 2p_{t+1}^k p_t^k | \mathcal{H}_t))$$
$$= \mathbb{E}((p_{t+1}^k)^2) - \mathbb{E}((p_t^k)^2).$$

So

$$\mathbb{E}\left(\sum_{t=0}^{T-1} (p_{t+1}^k - p_t^k)^2\right) = \mathbb{E}\left((p_T^k)^2\right) - (p^k)^2 \leqslant p^k(1 - p^k).$$

By the Cauchy–Schwarz inequality, we also have

$$\mathbb{E}\left(\frac{1}{T} \sum_{t=0}^{T-1} |p_{t+1}^k - p_t^k|\right) \leqslant \sqrt{\frac{1}{T} \mathbb{E}\left(\sum_{t=0}^{T-1} (p_{t+1}^k - p_t^k)^2\right)}$$

for each k, and the result follows. □

As soon as player 1 uses a strategy which depends on the selected state, the martingale of a posteriori will move and player 2 will have learnt something about the state. This is the dilemma of the informed player: he cannot use the information on the state without revealing information. For h_t in $(I \times J)^t$, $\sigma_{t+1}^1(k, h_t)$ is the mixed action in $\Delta(I)$ played by player 1 at stage $t + 1$ if the state is k and h_t has previously occurred, and we write $\bar{\sigma}_{t+1}^1(h_t)$ for the law of the action of player 1 of stage $t + 1$ after h_t: $\bar{\sigma}_{t+1}^1(h_t) = \sum_{k \in K} p_t^k(h_t) \sigma_{t+1}^1(k, h_t) \in \Delta(I)$. $\bar{\sigma}_{t+1}^1(h_t)$ can be seen as the average action played by player 1 after h_t, and will be used as a non-revealing approximation for $(\sigma_{t+1}^1(k, h_t))_k$. The next lemma precisely links the variation of the martingale $(p_t(\sigma))_{t \geqslant 0}$, i.e. the information revealed by player 1, and the dependence of player 1's action on the selected state, i.e. the information used by player 1.

Lemma 8.6.8

$$\forall t \geqslant 0, \forall h_t \in (I \times J)^t, \quad \mathbb{E}(\|p_{t+1} - p_t\| \,|h_t) = \mathbb{E}\left(\left\|\sigma_{t+1}^{\tilde{k}}(h_t) - \bar{\sigma}_{t+1}(h_t)\right\| \,|h_t\right).$$

Proof Fix $t \geqslant 0$ and h_t in $(I \times J)^t$ s.t. $\mathbb{P}_{p,\sigma}(h_t) > 0$. For (i_{t+1}, j_{t+1}) in $I \times J$, we have

$$
\begin{aligned}
p_{t+1}^k(h_t, i_{t+1}, j_{t+1}) &= \mathbb{P}(\tilde{k} = k|h_t, i_{t+1}) \\
&= \frac{\mathbb{P}(\tilde{k} = k|h_t)\mathbb{P}(i_{t+1}|k, h_t)}{\mathbb{P}(i_{t+1}|h_t)} \\
&= \frac{p_t^k(h_t)\sigma_{t+1}^1(k, h_t)(i_{t+1})}{\bar{\sigma}_{t+1}^1(h_t)(i_{t+1})}.
\end{aligned}
$$

Consequently,

$$
\begin{aligned}
\mathbb{E}(\|p_{t+1} - p_t\| \,|h_t) &= \sum_{i_{t+1} \in I} \bar{\sigma}_{t+1}^1(h_t)(i_{t+1}) \sum_{k \in K} |p_{t+1}^k(h_t, i_{t+1}) - p_t^k(h_t)|. \\
&= \sum_{i_{t+1} \in I} \sum_{k \in K} |p_t^k(h_t)\sigma_{t+1}^1(k, h_t)(i_{t+1}) - \bar{\sigma}_{t+1}^1(h_t)(i_{t+1})p_t^k(h_t)| \\
&= \sum_{k \in K} p_t^k(h_t)\|\sigma_{t+1}^1(k, h_t) - \bar{\sigma}_{t+1}^1(h_t)\| \\
&= \mathbb{E}\left(\|\sigma_{t+1}^1(\tilde{k}, h_t) - \bar{\sigma}_{t+1}^1(h_t)\| \,|h_t\right). \qquad \Box
\end{aligned}
$$

We can now control payoffs. For $t \geqslant 0$ and h_t in $(I \times J)^t$:

$$
\begin{aligned}
\mathbb{E}\left(G^{\tilde{k}}(\tilde{i}_{t+1}, \tilde{j}_{t+1})|h_t\right) &= \sum_{k \in K} p_t^k(h_t)G^k(\sigma_{t+1}^1(k, h_t), \sigma_{t+1}^2(h_t)) \\
&\leqslant \sum_{k \in K} p_t^k(h_t)G^k(\bar{\sigma}_{t+1}^1(h_t), \sigma_{t+1}^2(h_t)) \\
&\quad + M \sum_{k \in K} p_t^k(h_t)\|\sigma_{t+1}^1(k, h_t) - \bar{\sigma}_{t+1}^1(h_t)\| \\
&\leqslant u(p_t(h_t)) + M \sum_{k \in K} p_t^k(h_t)\|\sigma_{t+1}^1(k, h_t) - \bar{\sigma}_{t+1}^1(h_t)\|,
\end{aligned}
$$

where $u(p_t(h_t))$ comes from the definition of σ^2. By Lemma 8.6.8, we get

$$\mathbb{E}\left(G^{\tilde{k}}(\tilde{i}_{t+1}, \tilde{j}_{t+1})|h_t\right) \leqslant u(p_t(h_t)) + M\mathbb{E}(\|p_{t+1} - p_t\| \,|h_t).$$

Applying Jensen's inequality yields

$$\mathbb{E}\left(G^{\tilde{k}}(\tilde{\imath}_{t+1}, \tilde{\jmath}_{t+1})\right) \leqslant \operatorname{cav} u(p) + M\mathbb{E}\left(\|p_{t+1} - p_t\|\right).$$

We now apply Lemma 8.6.7 and obtain:

$$\gamma_T^{1,p}(\sigma^1, \sigma^2) = \mathbb{E}\left(\frac{1}{T}\sum_{t=0}^{T-1} G^{\tilde{k}}(\tilde{\imath}_{t+1}, \tilde{\jmath}_{t+1})\right)$$

$$\leqslant \operatorname{cav} u(p) + \frac{M}{\sqrt{T}} \sum_{k\in K} \sqrt{p^k(1-p^k)}.$$

Since this is true for any strategy σ^1 of player 1, we obtain

$$\forall p \in \Delta(K), \forall T \geqslant 1, \quad v_T(p) \leqslant \operatorname{cav} u(p) + \frac{M\sum_{k\in K}\sqrt{p^k(1-p^k)}}{\sqrt{T}}. \tag{8.7}$$

Finally, Theorem 8.6.6 follows from (8.4), (8.6) and (8.7). □

Aumann and Maschler also showed that for this class of games, $\lim_{\lambda\to 0} v_\lambda(p) = \operatorname{cav} u(p)$, that the uniform value exists and both players have optimal strategies, i.e.:

$$\exists \sigma^1, \forall \varepsilon > 0, \quad \exists T_0, \forall T \geqslant T_0, \quad \forall \sigma^2, \quad \gamma_T^1(\sigma^1, \sigma^2) \geqslant \operatorname{cav} u(p) - \varepsilon,$$

$$\exists \sigma^2, \forall \varepsilon > 0, \quad \exists T_0, \forall T \geqslant T_0, \quad \forall \sigma^1, \quad \gamma_T^1(\sigma^1, \sigma^2) \leqslant \operatorname{cav} u(p) + \varepsilon.$$

Theorem 8.6.6 and the above results have been generalized in many ways since the 1960s (see in particular the books [137, 203]). For zero-sum repeated games with incomplete information, the case of incomplete information on both sides has been extensively investigated, e.g.: convergence of $(v_T)_T$ and $(v_\lambda)_\lambda$ to the unique solution of a system of functional equations [138, 140], speed of convergence [41, 42, 139], lack of information on one and a half sides [205], the splitting game [115, 116], extension of the convergence to games where the private states of the players evolve according to exogenous and independent Markov chains [76]. General results also exist in the one-player case of dynamic programming or Markov decision problems with imperfect signals on the state [166, 171, 179], with extensions to zero-sum repeated games where one player is informed of the state and controls the transitions [167]. For two-player general-sum repeated games with lack of information on one side, the existence of uniform equilibrium was proved in Sorin [197] for two states and generalized by Simon et al. [190] for any number of states (extended to state independent signals in [164]); and a characterization of equilibrium payoffs has been given by Hart [94] (see also [8]).

Let us finally mention the existence, in the class of two-player *hidden stochastic games* (stochastic games with public signals on the unobserved state), of robust counter-examples for the convergence of the T-stage or λ-discounted equilibrium payoffs [232] for the values in the zero-sum case, Renault–Ziliotto [172] for the equilibrium payoff sets in the non-zero-sum case.

8.7 Exercises

Exercise 1. Feasible and IR Payoffs
Compute the set of feasible and individually rational payoffs in the following games:

(1)

	L	R
T	$(1, 1)$	$(3, 0)$
B	$(0, 3)$	$(0, 0)$

(2)

	L	R
T	$(1, 0)$	$(0, 1)$
B	$(0, 0)$	$(1, 1)$

(3)

	L	R
T	$(1, -1)$	$(-1, 1)$
B	$(-1, 1)$	$(1, -1)$

(4)

	L	R			L	R
T	$(0, 0, 0)$	$(0, 1, 0)$		T	$(0, 0, 1)$	$(1, 0, 0)$
B	$(1, 0, 0)$	$(0, 0, 1)$		B	$(0, 1, 0)$	$(0, 0, 0)$

$$W \qquad\qquad\qquad\qquad E$$

Exercise 2. Zero-sum repeated games
Let G be a finite two-player *zero-sum* game. What are the equilibrium payoffs of the finitely repeated game G_T, of the discounted game G_λ, and of the uniform game G_∞?

Exercise 3. Cesaro and Abel evaluations
Prove Lemma 8.3.5.

Exercise 4. (Sorin [198])
Let G be a finite two-player game such that $E_1 = \{(v^1, v^2)\}$, i.e. there is a unique Nash equilibrium payoff of G where both players receive their independent minmax. Show that $E_T = \{(v^1, v^2)\}$ for each T.

Exercise 5. Minmax not feasible
Find a 2-player finite game G where the vector of independent minmax $v = (v^i)_{i \in N}$ is not feasible.

Exercise 6. Repetition and cooperation
Let G be the following strategic game:

	L	R
T	$(0, 0)$	$(0, 2)$
B	$(2, 0)$	$(0, 0)$

(1) Compute the mixed Nash equilibrium payoffs, and the correlated equilibrium payoffs of G.

(2) We assume that G is played twice, and after the first stage the actions actually played are publicly announced. We also assume that the payoffs of the players are just given by the actions of stage 2. This defines a new game Γ. Show that $(1, 1)$ is an equilibrium payoff of Γ.

Exercise 7. Prisoner's dilemma with a blind player

Consider a repeated prisoner's dilemma where at the end of every stage, player 2 observes the action of player 1 but player 1 observes nothing of the past actions of player 2.

$$
\begin{array}{c|c|c|}
 & C^2 & D^2 \\
\hline
C^1 & (3, 3) & (0, 4) \\
\hline
D^1 & (4, 0) & (1, 1) \\
\hline
\end{array}
$$

Given $\lambda \in [0, 1)$, compute the set of equilibrium payoffs E_λ of the λ-discounted repeated game.

Exercise 8. Battle of the sexes in the dark

Consider the battle of sexes played offline (in the dark): the players do not observe the actions of the other player between the stages.

$$
\begin{array}{c|c|c|}
 & C^2 & D^2 \\
\hline
C^1 & (2, 1) & (0, 0) \\
\hline
D^1 & (0, 0) & (1, 2) \\
\hline
\end{array}
$$

Compute $\bigcup_{T \geq 1} E_T$, where E_T is the set of Nash equilibrium payoffs of the T-stage repeated game with average payoffs.

Exercise 9. Jointly rational payoffs

Consider the following three-player repeated game with signals.

$$
\begin{array}{c|c|c|}
 & L & R \\
\hline
T & (0, 0, 0) & (0, 1, 0) \\
\hline
B & (1, 0, 0) & (1, 1, 0) \\
\hline
\end{array}
\qquad
\begin{array}{c|c|c|}
 & L & R \\
\hline
T & (0, 1, 0) & (0, 1, 0) \\
\hline
B & (0, 1, 0) & (0, 1, 0) \\
\hline
\end{array}
$$

$$W \qquad\qquad\qquad M$$

$$
\begin{array}{c|c|c|}
 & L & R \\
\hline
T & (1, 0, 0) & (1, 0, 0) \\
\hline
B & (1, 0, 0) & (1, 0, 0) \\
\hline
\end{array}
$$

$$E$$

Player 1 chooses the row, player 2 chooses the column and player 3 chooses the matrix. At the end of every stage, players 1 and 2 perfectly observe all actions played, whereas player 3 observes the signal s if the action profile played by player 1 and player 2 is (T, L), and player 3 observes s' otherwise. Notice that the payoff of player 3 is identically 0.

(1) Compute the set of feasible and individually rational payoffs.
(2) Show that for every $T \geqslant 1$, $E_T \subset \{(u^1, u^2, 0) \in \mathbb{R}^3, u^1 + u^2 \geqslant 1\}$.
(3) Show that $E_T \xrightarrow[T \to \infty]{} \mathrm{co}\{(1, 0, 0), (0, 1, 0), (1, 1, 0)\}$.

Exercise 10. Subgame-perfect equilibrium payoffs in discounted games

Fix a standard repeated game $G = (N, (A^i)_{i \in N}, (g^i)_{i \in N})$ with discount λ in $(0, 1]$. A strategy of player i is a mapping σ^i from H to $\Delta(A^i)$, where $H = \bigcup_{t \geqslant 0} H_t$ is the set of possible histories (or information sets) of the game. A one-shot deviation of σ^i is a strategy τ^i of the same player which differs from σ^i at a single history in H, and we denote by $OS(\sigma^i)$ the set of one-shot deviations of σ^i. Recall that a strategy profile $\sigma = (\sigma^i)_{i \in N}$ is a subgame-perfect equilibrium of G_λ if for each history h in H, the continuation strategy profile $\sigma[h]$ is a Nash equilibrium of G_λ.

(1) *One-shot deviation principle.* Let $\sigma = (\sigma^i)_{i \in N}$ be a mixed strategy profile. Show that σ is a subgame-perfect equilibrium of G_λ if and only if there is no profitable one-shot deviation, i.e.

$$\forall i \in N, \forall h \in H, \forall \tau^i \in OS(\sigma^i), \ \gamma_\lambda^i(\sigma^{-i}[h], \tau^i[h]) \leqslant \gamma_\lambda^i(\sigma[h]).$$

(2) Write $R = \mathbb{R}^N$ for the space of payoff profiles. The vector payoff function is a map $g : A \to R$. Given any $u : A \to R$, we denote by $\mathrm{NE}(u)$, resp. $\mathrm{NEP}(u)$, the set of mixed Nash equilibria, resp. mixed Nash equilibrium payoffs, of the strategic game with set of players N, actions sets A^i for each player i, and vector payoff function u. We define the map Φ which associates to any non-empty compact F of R the non-empty compact subset of R:

$$\Phi(F) = \bigcup_{u : A \to F} \mathrm{NEP}(\lambda g + (1 - \lambda)u).$$

(a) Show that the set E'_λ of subgame-perfect equilibrium payoffs of G_λ is a fixed point of Φ.
(b) Show that E'_λ is the largest fixed point of Φ (Abreu et al. [1]).

Exercise 11. On ontinuation payoffs

Prove Lemma 8.5.7.

Exercise 12. Quitting Games

Fix a finite set of players $N = \{1, \ldots, n\}$ and for each non-empty subset S of N, a vector $r(S)$ in $\mathbb{R}^n$. The following stochastic game is called a *quitting game*: At each stage $1, \ldots, t, \ldots$, the players independently choose either to quit or continue. If everyone continues, the play goes to stage $t + 1$. If a subset S of players quits at stage t, each player i receives the payoff $r_i(S)$ and the game is over. If nobody ever quits, each player receives the payoff 0.

A mixed strategy of player i is a probability distribution x_i on $T = \{1, \ldots, t, \ldots\}$ $\cup \{+\infty\}$ (quitting at time $+\infty$ meaning never quitting), and as usual a mixed strategy profile $x = (x_j)_{j \in N}$ induces an expected payoff $g_i(x)$ for each player i.

Prove that a quitting game with two players has an equilibrium payoff: given $\varepsilon > 0$, show that there exists a ε-Nash equilibrium, i.e. a mixed strategy profile $x = (x_j)_{j \in N}$ such that

$$\forall i \in N, \forall y_i \in \Delta(T), \ g_i(y_i, x_{-i}) \leqslant g_i(x) + \varepsilon.$$

Remark The existence of equilibrium payoffs in quitting games is open for $n \geqslant 4$ (for $n = 3$, see Solan and Vieille [196]).

Chapter 9
Solutions to the Exercises

9.1 Hints for Chapter 1

Exercise 1

(1) The algorithm is well defined since if a given man has already proposed to every woman, every woman already had a man at home and thus has exactly one man at home, terminating the algorithm.

To show that it stops in at most n^2 days, consider the state variable giving for every man the number of women he has visited. If a man has visited all n women, the algorithm stops. So as long as the algorithm continues on day k, at the beginning of this day the state variable takes values in $\{0, \ldots, n-1\}^n$, and at least one man will visit a new woman on this day.

(2) First example: three stable matchings

$$((a, A), (b, B), (c, C)), \quad ((b, A), (c, B), (a, C)) \quad \text{and} \quad ((c, A), (a, B), (b, C)).$$

Second example: a unique stable matching $((c, A), (d, B), (a, C), (b, D))$.

(3) If a man prefers a woman to the one he is matched with, he made a proposal to her before; hence she cannot be matched with a man she ranks worse.

(4) Consider, for instance, four students A, B, C, D with D being the least popular for everyone, and A ranks B first, B ranks C first, and C ranks A first.

(5) Similar to the monogamy case with the following algorithm: each student first applies to his preferred school, then to his second best choice in case of refusal, etc. and a school with capacity q handles a waiting list composed of its preferred students (at most q) among those who have applied to it. The algorithm stops if each student is either in some waiting list, or has been rejected everywhere. The waiting lists of each school are then finally accepted.

Exercise 2

(1) Define $x^* = f^{-1}(1/2)$ and $y^* = g^{-1}(1/2)$. By stopping at x^* (if player 2 has not stopped before), player 1 can make sure to have at least $1/2$. Similarly by stopping

© Springer Nature Switzerland AG 2019
R. Laraki et al., *Mathematical Foundations of Game Theory*, Universitext,
https://doi.org/10.1007/978-3-030-26646-2_9

at y^*, player 2 guarantees $1/2$. Notice that player 1 (resp. player 2) should never stop before x^* (resp y^*).

(2) If $x^* \leqslant y^*$, it is reasonable to expect that both players will stop at y^*, with a payoff $f(y^*)$ for player 1 and $1/2$ for player 2. If $x^* > y^*$, one can expect player 1 to stop at x^*, and player 2 slightly before x^* (see the notion of ε-equilibrium later in Chap. 4).

(3) No clear answer. One possibility for player 1 (resp. player 2) is to stop at x^* (resp. y^*). This is a prudent strategy guaranteeing $1/2$ in all cases (but it might be too soon).

(4) No. Think for instance of a cake with a strawberry at the left and a chocolate piece at the right, with player 1 loving strawberries and player 2 loving chocolate.

Exercise 3

(1) $(x_t)_t$ is decreasing positive, and $x_t \xrightarrow[t\to\infty]{} 0$.

(2) $u(V, x) = x$ and $u(B, x) = x - 1/2$ for all x.

(3) One can show that $(x_t - 1/3)^2 \xrightarrow[t\to\infty]{} 0$.

(4) and (5) Draw pictures.

Exercise 4

Fix a player i, denote by p_i his bid, by $p_{-i} = (p_j)_{j\neq i}$ the vector of bids of the other players, and define $p^* = \max_{j\neq i} p_j$. Player i's utility is 0 if $p_i < p^*$, it is $v_i - p^*$ if $p_i > p^*$, and it is either 0 or $v_i - p^*$ if $p_i = p^*$. Let us show that $u_i(v_i, p_{-i}) \geqslant u_i(p_i, p_{-i})$ for all p_i and p_{-i}.

If $v_i > p^*$, then $u_i(v_i, p_{-i}) = v_i - p^* > 0$ and $v_i - p^* \geqslant u_i(p_i, p_{-i})$.

If $v_i = p^*$, then $u_i(v_i, p_{-i}) = 0 = u_i(p_i, p_{-i})$.

If $v_i < p^*$, then $u_i(v_i, p_{-i}) = 0 \geqslant u_i(p_i, p_{-i})$.

9.2 Hints for Chapter 2

Exercise 1

(1.b) This implies $\lambda_0 = \mu_0$.

(1.c) The configuration A', B' is more favorable to player 1 (fewer constraints) hence, since its value exists by induction, we have $v' \geqslant \mu_0 > \lambda_0$.

Then for $\alpha \in (0, 1)$ small enough $(\alpha s' + (1 - \alpha)s_0)(A - \lambda_0 B) \gg 0$.

This contradicts the definition of λ_0.

(2) Consider $A = \text{Id}$. There exists s and t such that $s \geqslant v\, sB$ and $t \leqslant v\, Bt$.

$t \leqslant v\, Bt$ implies $v > 0$.

$s \geqslant v\, sB$ implies that s has full support. Thus by complementarity $t = v\, Bt$ hence t also has full support.

So again by duality $s = v\, sB$.

Exercise 2

Let $\alpha = \max_{(x,y)} \min_z g(x, y, z)$.

Thus $\alpha = \max_{(x,y) \in [0,1]^2} \min\{xy, (1-x)(1-y)\}$. Let $a = xy$ and $b = (1-x)$ $(1-y)$. Since $ab = x(1-x)y(1-y) \leqslant 1/16$, we have $\min\{a, b\} \leqslant 1/4$, and $\alpha \leqslant 1/4$. Finally, $x = y = 1/2$ gives $\alpha = 1/4$.

Write $\beta = \min_z \max_{(x,y)} g(x, y, z)$. Hence $\beta = \min_{z \in [0,1]} \max\{z, 1-z\} = 1/2$. We obtain $\alpha < \beta$.

Remark: The duality gap is due to the lack of convexity of the strategy set of players 1 and 2 together. Using correlation, i.e. allowing players 1 and 2 to play a probability in $\Delta(\{T, B\} \times \{L, R\})$ instead of $\Delta(\{T, B\}) \otimes \Delta(\{L, R\})$ would increase α to 1/2.

Exercise 3

(1) Let $v = \limsup v_n$ and $\varepsilon > 0$. Choose n such that $v_n \geqslant v - \varepsilon$ and $|x - y| \leqslant \frac{1}{2^{n+1}}$ implies $|f(x) - f(y)| \leqslant \varepsilon$ (f is uniformly continuous on the square).

Let σ_n be optimal for player 1 in G_n and $t \in [0, 1]$. There exists a $j_n = 2^n t_n$ in Y_n such that $|t - t_n| \leqslant \frac{1}{2^{n+1}}$. Then

$$f(\sigma_n, t) = f(\sigma_n, t_n) + \int_s f(s, t) - f(s, t_n) d\sigma_n(s) \geqslant v - \varepsilon - \varepsilon.$$

Thus σ_n guarantees $v - 2\varepsilon$ in G.

A dual result then implies that the sequence v_n converges to v, which is the value of G.

(2) For each n, consider the optimal strategy σ_n of player 1 in G_n, and the associated repartition function $F_n : \mathbb{R} \longrightarrow [0, 1]$. The corresponding set of functions is sequentially weakly compact. (Construct by a diagonal procedure a subsequence converging on rational points, then choose the induced right continuous extension.)

Exercise 4

(1) Let $y_n = \Pi_C(\bar{x}_n)$. Then

$$d_{n+1}^2 \leqslant \|\bar{x}_{n+1} - y_n\|^2 = \|\bar{x}_n - y_n\|^2 + \|\bar{x}_{n+1} - \bar{x}_n\|^2 + 2\langle \bar{x}_{n+1} - \bar{x}_n, \bar{x}_n - y_n \rangle.$$

Decompose

$$\langle \bar{x}_{n+1} - \bar{x}_n, \bar{x}_n - y_n \rangle = \left(\frac{1}{n+1}\right) \langle x_{n+1} - \bar{x}_n, \bar{x}_n - y_n \rangle$$
$$= \left(\frac{1}{n+1}\right) (\langle x_{n+1} - y_n, \bar{x}_n - y_n \rangle - \|\bar{x}_n - y_n\|^2).$$

Using the hypothesis we obtain

$$d_{n+1}^2 \leqslant \left(1 - \frac{2}{n+1}\right) d_n^2 + \left(\frac{1}{n+1}\right)^2 \|x_{n+1} - \bar{x}_n\|^2.$$

From

$$\|x_{n+1} - \bar{x}_n\|^2 \leqslant 2\|x_{n+1}\|^2 + 2\|\bar{x}_n\|^2 \leqslant 4M^2,$$

we deduce

$$d_{n+1}^2 \leqslant \left(\frac{n-1}{n+1}\right) d_n^2 + \left(\frac{1}{n+1}\right)^2 4M^2.$$

Thus by induction

$$d_n \leqslant \frac{2M}{\sqrt{n}}.$$

(2.a) By construction: $\langle s_{n+1} A, t_{n+1} \rangle \geqslant 0$, thus

$$\langle x_{n+1}, \bar{x}_n^+ - \bar{x}_n \rangle \geqslant 0.$$

Since $\langle \bar{x}_n^+, \bar{x}_n - \bar{x}_n^+ \rangle = 0$ we get

$$\langle x_{n+1} - \bar{x}_n^+, \bar{x}_n - \bar{x}_n^+ \rangle \leqslant 0$$

and recall that $\bar{x}_n^+ = \Pi_C(\bar{x}_n)$.

(2.b) Consider the empirical frequencies arising in $\bar{x}_n$ as a mixed strategy of player 1 and use compactness of $\Delta(I)$.

Exercise 5

Strategy a_2 is strictly dominated by $0.51\, a_3 + 0.49\, a_1$. Hence the reduced game is

	b_1	b_2
a_1	3	−1
a_3	−2	1

with unique optimal strategies $(3/7, 4/7)$; $(2/7, 5/7)$ and value $1/7$.

Exercise 6

The strategy that plays i with probability $\dfrac{\frac{1}{a_i}}{\sum_{j=1}^{n} \frac{1}{a_j}}$ is equalizing with payoff $v = \dfrac{1}{\sum_{j=1}^{n} \frac{1}{a_j}}$ whatever the action of player 2. By symmetry v is thus the value of the game.

Exercise 7

(1)

$$0 = \langle uA, x \rangle = uAx = \langle u, Ax \rangle \geqslant \langle u, b \rangle > 0,$$

a contradiction.

(2) For $n = 0$, the matrix A is 0. S empty means that b has a positive component, say $b_i > 0$. Then the unit vector e^i belongs to T.

(3.a) Simply complete $\bar{u}$ with zero components.

(3.b) The product with the last line $j = n + 1$ from A gives

$$\sum_\ell u_\ell A_{\ell,n+1} + \sum_k u_k A_{k,n+1} = \sum_{\ell,k} v_{\ell,k} - \sum_{k,\ell} v_{\ell,k} = 0$$

and the other constraints follow from the definition of v.

Exercise 8

If $t > 0$, let $x = \frac{r-q}{t}$.

If $t = 0$, let x' with $Ax' \geqslant b$. Then for $a > 0$ large enough, $x = x' + a(r - q)$ satisfies

$$Ax = Ax' + aA(r - q) \geqslant b, \qquad \langle c, x \rangle = \langle c, x' \rangle + a\langle c, r - q \rangle < d.$$

Exercise 9

(2) Note that for any pair of feasible points (x, u) we have

$$cx \geqslant uAx \geqslant ub.$$

Exercise 10

(2.a) One can write z as $(x, y, t) \in \mathbb{R}_+^I \times \mathbb{R}_+^J \times \mathbb{R}_+$, check that $x > 0$, $y > 0$, $t > 0$ and use complementarity.

Exercise 11

(1) Fix j and assume that there is no optimal solution x with $x_j > 0$. Then denoting by k the common value of both programs we obtain

$$Ax \geqslant b, \quad x \geqslant 0, \quad -\langle c, x \rangle \geqslant k \quad \Rightarrow -\langle e_j, x \rangle \geqslant 0,$$

thus using Farkas' Lemma, there exist (u, v, w) with

$$(u, v, w) \geqslant 0, \quad uA + v - wc = e_j, \quad \langle u, b \rangle - wk \geqslant 0.$$

If $w > 0$, take $u' = u/w$, which satisfies $c_j - u'A_j > 0$.

If $w = 0$, take some optimal u^*, then $u' = u + u^*$, and the same property holds.

(2) Thus for each $i \in I$, there exists a pair of optimal solutions $(x(i), u(i))$ satisfying the strong complementarity condition for this index, and similarly for any $j \in J$. It is then clear that the barycenter

$$(\bar{x}, \bar{u}) = \frac{1}{(|I| + |J|)} \left[\sum_i (x(i), u(i)) + \sum_j (x(j), u(j)) \right]$$

will satisfy the requested requirement.

9.3   Hints for Chapter 3

Exercise 1

(1) The payoff of player 1 is given by

$$
g(x, y) = \begin{cases}
p_1(x) + (1 - p_1(x))(-1) = 2p_1(x) - 1 & \text{if } x < y \\
p_1(x)(1 - p_2(x)) + p_2(x)(1 - p_1(x))(-1) & \\
\qquad\qquad\qquad\qquad = p_1(x) - p_2(x) & \text{if } x = y \\
(-1)p_2(y) + (1 - p_2(y)) = 1 - 2p_2(y) & \text{if } x > y.
\end{cases}
$$

Let $t_0 \in (0, 1)$ be such that $p_1(t_0) + p_2(t_0) = 1$. Then $g(t_0, t_0) = 2p_1(t_0) - 1$, $g(t_0, y) \geqslant 2p_1(t_0) - 1$ for each y, and $g(x, t_0) \leqslant 2p_1(t_0) - 1$ for each x. So (t_0, t_0) is a saddle-point. The optimal strategy of a player is to shoot at the specific time where the probability of killing the adversary equals the probability of surviving while being shot at. At the optimum, players shoot simultaneously.

(2) Let us prove by induction that for each K, the following property (H_K) holds:

(H_K): When there are (m, n) bullets with $m + n \leqslant K$, the value exists and is $\frac{m-n}{m+n}$, and an optimal strategy of the player having $\max(m, n)$ bullets is to shoot* for the first time at $t = \frac{1}{m+n}$.

(H_1) is true. Assume (H_K) holds for some $K \geqslant 1$ and suppose that there are $(m', n) = (m + 1, n)$ bullets. Without loss of generality we assume $n > 0$ and $m' > 0$. We distinguish three cases, depending on which player has more bullets.

Case 1: $m' > n > 0$. Define the following strategy σ of player 1: in case of silence shoot for the first time at $\frac{1}{m+n+1}$, then if the shooting has failed, play optimally in the game (m, n); if player 2 shoots and fails before time $\frac{1}{m+n+1}$, player 1 plays optimally in the game $(m + 1, n - 1)$ (if both players shoot for the first time and fail at $\frac{1}{m+n+1}$, play optimally in the game $(m, n - 1)$).

If player 2 only shoots after $\frac{1}{m+n+1}$, the payoff of player 1 using σ is at least

$$
\frac{1}{m+n+1} + \frac{m+n}{m+n+1}\,\frac{m-n}{m+n} = \frac{m+1-n}{m+n+1}.
$$

If player 2 shoots for the first time at $\frac{1-\varepsilon}{m+n+1} < \frac{1}{m+n+1}$, the payoff of player 1 using σ is at least

$$
\frac{1-\varepsilon}{m+n+1}(-1) + \frac{m+n+\varepsilon}{m+n+1}\,\frac{m+1-(n-1)}{m+n} > \frac{m+1-n+\varepsilon}{m+n+1}.
$$

If player 2 shoots for the first time exactly at $\frac{1}{m+n+1}$, the payoff of player 1 using σ is at least

$$
0 + \left(\frac{m+n}{m+n+1}\right)^2 \frac{m-n+1}{m+n-1} > \frac{m+1-n}{m+n+1}.
$$

Consequently, σ guarantees $\frac{m+1-n}{m+n+1}$ to player 1.

Define now, for small $\varepsilon > 0$, the random strategy τ of player 2 who, in case of silence, shoots for the first time according to the uniform law on the interval $[\frac{1}{m+n+1}, \frac{1+\varepsilon}{m+n+1}]$, and plays optimally as soon as one bullet has been used. If player 1 shoots for the first time before player 2, his payoff is at most $\frac{1+\varepsilon}{m+n+1} + \frac{m+n-\varepsilon}{m+n+1}\frac{m-n}{m+n} \leqslant \frac{m+1-n+\varepsilon}{m+n+1}$. If player 1 shoots for the first time before player 2, his payoff is at most $\frac{1}{m+n+1}(-1) + \frac{m+n}{m+n+1}\frac{m+2-n}{m+n} = \frac{m+1-n}{m+n+1}$. So player 1 also guarantees $\frac{m+1-n}{m+n+1}$ (up to ε for each ε), and this is the value of the game: the conclusion of (H_{K+1}) holds (notice that player 2 has no ε-optimal pure strategy here).

Case 2: $m' = n$. Define σ (resp. τ) the strategy of player 1 (resp. player 2) who, in case of silence, shoots for the first time at time $\frac{1}{m+n+1}$, and plays optimally as soon as a bullet has been used.

If player 1 shoots for the first time before $\frac{1}{m+n+1}$, let's say at $t = \frac{1-\varepsilon}{m+n+1}$ with $\varepsilon > 0$, his payoff against τ is at most $g(t, \tau) = \frac{1-\varepsilon}{m+n+1} + \frac{m+n+\varepsilon}{m+n+1}\frac{(-1)}{m+n} < 0$.

If player 1 shoots for the first time at $t > \frac{1}{m+n+1}$, his payoff against τ is at most $g(t, \tau) = \frac{-1}{m+n+1} + \frac{m+n}{m+n+1}\frac{1}{m+n} = 0$.

So $g(t, \tau) \leqslant 0$ for each t, and by symmetry (σ, τ) is a saddle-point of the game. Here again, the conclusions of (H_{K+1}) hold.

Case 3: $m' < n$. This is similar to case 1.

(3.a) The payoff of player 1 is now

$$g(x, y) = \begin{cases} x + (1-x)y(-1) = -(1-x)y + x & \text{if } x < y \\ x(1-x) + x(1-x)(-1) = 0 & \text{if } x = y \\ (-1)y + (1-y)x = -(1+x)y + x & \text{if } x > y \end{cases}$$

If $y > 0$, playing x just before y gives $\sup_x g(x, y) \geqslant y^2$, and if $y < 1$, we have $g(1, y) = 1 - 2y$, so $\inf_y \sup_x g(x, y) > 0$. Yet the game is symmetric, hence if it exists the value can only be 0. There is no value in pure strategies.

(3.b) Assume now that player 1 has a mixed strategy σ with density f on $[\alpha, 1]$, with $\alpha > 0$ and f differentiable, which guarantees a non-negative payoff. Then, by symmetry, σ also guarantees 0 for player 2, and we expect that $g(\sigma, y) = 0$ for each $y \geqslant \alpha$, and so

$$-y + (1+y)\int_\alpha^y xf(x)dx + (1-y)\int_y^1 xf(x)dx = 0.$$

Differentiating twice, we find $yf'(y) + 3f(y) = 0$. We deduce that $f(y) = Cy^{-3} 1_{y \geqslant a}$, for some constant C. The above equation gives $C = 1/4$, and $a = 1/3$. It is finally enough to check the above computations with the strategy σ now well defined.

(4) The payoff function is the following

$$g(x, y) = \begin{cases} x - y + xy & \text{if } x < y \\ 0 & \text{if } x = y \\ 1 - 2y & \text{if } x > y, \end{cases}$$

and one verifies that the given strategies do form a saddle-point.

(5) It is not possible to proceed by induction in a silent duel, since once a bullet has been used only the player who shot is aware of it.

Exercise 2

(1) $\sup_s \inf_t f(s, t) = -1$, $\inf_t \sup_s f(s, t) = 0$.

One can check that f is quasi-concave in s and quasi-convex and l.s.c. in t. It is also u.s.c. in s for each t, except for $t = 1$.

The value in mixed strategies exists and is $-\frac{1}{2}$. It is enough for player 1 to play uniformly on $[0, 1]$ (or to play 0 with probability $\frac{1}{2}$ and 1 with probability $\frac{1}{2}$).

Exercise 3

(1) Write $v = \inf_n v_n = \lim_n v_n$. For n such that $v_n \leqslant v + 1/n$, consider a strategy t_n which is $1/n$-optimal for player 2 in G_n. For each s in S, $f(s, t_n) \leqslant f_n(s, t_n) \leqslant v_n + 1/n \leqslant v + 2/n$. We now define for each n:

$$A_n = \{s \in S, f_n(s, t) \geqslant v - \tfrac{1}{n} \; \forall t \in T\}.$$

A_n is a decreasing sequence of non-empty compact sets, hence $\bigcap_n A_n \neq \varnothing$. Considering an element in $\bigcap_n A_n$ gives an optimal strategy of player 1 in G.

(2) In both cases, for each n we have $v_n = 1$ and $v = 0$. In the first case S is not compact, and in the second $(f_n)_n$ is not monotonic.

Exercise 4

Simply write $\mathbb{E}$ for $\mathbb{E}_{\sigma,\tau}$.

(1) Fix $n \geqslant 1$ and h_n in $\mathcal{H}_n$. At stage $n + 1$, player 1 plays $s_{n+1} \in \Delta(I)$ such that for each t, $\langle s_{n+1} A t - \Pi_C(\bar{x}_n), \bar{x}_n - \Pi_C(\bar{x}_n) \rangle \leqslant 0$.

We have

$$d_{n+1}^2 \leqslant \|\bar{x}_{n+1} - \pi_C(\bar{x}_n)\|^2,$$

$$\leqslant \left\| \frac{1}{n+1} \sum_{t=1}^{n+1} x_t - \pi_C(\bar{x}_n) \right\|^2,$$

$$\leqslant \left\| \frac{1}{n+1}(x_{n+1} - \pi_C(\bar{x}_n)) + \frac{n}{n+1}(\bar{x}_n - \pi_C(\bar{x}_n)) \right\|^2,$$

$$\leqslant \left(\frac{1}{n+1} \right)^2 \|x_{n+1} - \pi_C(\bar{x}_n)\|^2 + \left(\frac{n}{n+1} \right)^2 d_n^2$$

$$+ \frac{2n}{(n+1)^2} \langle x_{n+1} - \pi_C(\bar{x}_n), \bar{x}_n - \pi_C(\bar{x}_n) \rangle.$$

By assumption, the expectation of the above scalar product is non-positive, so

$$\mathbb{E}\left(d_{n+1}^2|h_n\right) \leqslant \frac{1}{(n+1)^2}\mathbb{E}\left(\|x_{n+1} - \pi_C(\overline{x}_n)\|^2|h_n\right) + \left(\frac{n}{n+1}\right)^2 d_n{}^2.$$

(2) $\mathbb{E}\left(\langle x_{n+1} - \Pi_C(\overline{x}_n), \overline{x}_n - \Pi_C(\overline{x}_n)\rangle |h_n\right) \leqslant 0$, so

$$\mathbb{E}\left(\|x_{n+1} - \pi_C(\overline{x}_n)\|^2|h_n\right) \leqslant \mathbb{E}\left(\|x_{n+1} - \overline{x}_n\|^2|h_n\right) \leqslant (2\|A\|_\infty)^2.$$

(3) We obtain

$$\mathbb{E}\left(d_{n+1}^2|h_n\right) \leqslant \left(\frac{n}{n+1}\right)^2 d_n{}^2 + \left(\frac{1}{n+1}\right)^2 4\|A\|_\infty^2. \tag{9.1}$$

Taking expectation leads to: $\forall n \geqslant 1,\ \mathbb{E}\left(d_{n+1}^2\right) \leqslant \left(\frac{n}{n+1}\right)^2\mathbb{E}(d_n{}^2) + \left(\frac{1}{n+1}\right)^2 4\|A\|_\infty^2$. Then by induction, for each $n \geqslant 1$ we have $\mathbb{E}(d_n^2) \leqslant \frac{4\|A\|_\infty^2}{n}$, finally $\mathbb{E}(d_n) \leqslant \frac{2\|A\|_\infty}{\sqrt{n}}$.

(4) Inequality (9.1) yields: $\mathbb{E}(e_{n+1}|h_n) \leqslant e_n$, so (e_n) is a non-negative super-martingale with expectation converging to 0. So $e_n \longrightarrow_{n\to\infty} 0\ \mathbb{P}_{\sigma,\tau}$-a.s., and finally $d_n \longrightarrow_{n\to\infty} 0\ \mathbb{P}_{\sigma,\tau}$-a.s.

Exercise 5

Write $x = \sup_{s\in S}\inf_{t\in T} f(s,t)$ and $y = \inf_{\beta\in B}\sup_{s\in S} f(s,\beta(s))$. $x \leqslant y$ is clear. We show $x \geqslant y$ (using the axiom of choice). Fix $\varepsilon > 0$ and define, for each s in S, $\beta(s) \in T$ such that $f(s, \beta(s)) \leqslant \inf_t f(s,t) + \varepsilon$. Then $\sup_{s\in S} f(s,\beta(s)) \leqslant x + \varepsilon$.

Exercise 6

(3) Fubini's theorem does not apply, so $f(\sigma,\tau)$ and the mixed extension of the game is not well defined.

Exercise 7

(1) Let μ be the uniform distribution on $[0, 1/2]$, and δ_1 the Dirac measure on 1. The strategy $1/3\ \mu + 2/3\ \delta_1$ guarantees $1/3$ to player 1.

Suppose σ in $\Delta(S)$ guarantees a payoff not lower than $1/3$ to player 1: $f(\sigma,t) \geqslant 1/3$, $\forall t$.

Considering $t = 1$ implies $\sigma([0, 1/2[) - \sigma(]1/2, 1[) \geqslant 1/3$.

$t = \frac{1}{2}^-$ gives $-\sigma([0, 1/2[) + \sigma([1/2, 1]) \geqslant 1/3$.

Adding the inequalities: $\sigma(\{1/2\}) + \sigma(\{1\}) \geqslant 2/3$. Since $\sigma([0, 1/2[) \geqslant 1/3$, we get $\sigma([0, 1/2[) = 1/3$ and $\sigma(\{1/2\}) + \sigma(\{1\}) = 2/3$. So $f(\sigma, t = 1) = 1/3$, and σ does not guarantee more than $1/3$.

Conclusion: $\sup_\sigma \inf_\tau f(\sigma,\tau) = 1/3$.

(2) Let now τ in $\Delta(T)$ be such that $\inf_s f(s,\tau) \leqslant 1/3 + \varepsilon$.

$s = 1$ implies $\tau(\{1\}) \geqslant 2/3 - \varepsilon$.

$s = 0$ implies $\tau(]1/2, 1]) - \tau(]0, 1/2[) \leqslant 1/3 + \varepsilon$, so $\tau(]0, 1/2[) \geqslant 2/3 - \varepsilon - (1/3 + \varepsilon) = 1/3 - 2\varepsilon$.

$s = \frac{1}{2}^-$ gives $\tau(\{1\}) - \tau([1/2, 1]) + \tau([0, 1/2[) \leqslant 1/3 + \varepsilon$.

Hence $-1 + 2\tau(\{1\}) + 2\tau([0, 1/2[) \leqslant 1/3 + \varepsilon$, and $\tau(\{1\}) + \tau([0, 1/2[) \leqslant 2/3 + \varepsilon/2$.

This gives $1 - 3\varepsilon \leqslant 2/3 + \varepsilon/2$, so $\varepsilon \geqslant 2/21$, and $1/3 + \varepsilon \geqslant 3/7$. Hence $\inf_\tau \sup_\sigma$ $f(\sigma, \tau) \geqslant 3/7$, and G has no value.

Remark: one can show that $\inf_\tau \sup_\sigma f(\sigma, \tau) = 3/7$, by considering $\tau = 4/7\delta_1 + 2/7\delta_{1/2} + 1/7\delta_{1/4}$.

9.4 Hints for Chapter 4

Exercise 1

(1) The first game has three Nash equilibria: (T, R), (B, L) and $(\frac{2}{3}T + \frac{1}{3}B, \frac{2}{3}L + \frac{1}{3}R)$. To find them, we can explore all possible supports. Suppose $(xT + (1 - x)R, yB + (1 - y)L)$ is a Nash equilibrium, where x and y are in $[0, 1]$:

Case 1: $x = 1$ (player 1 plays T) $\Rightarrow y = 0$ (the best response of player 2 is R) $\Rightarrow x = 1$ (the best response of player 1 is T), as it should be. We have found the equilibrium (T, R).

Case 2: $x = 0 \Rightarrow y = 1 \Rightarrow x = 0$ (as it should be). We have found the equilibrium (B, L).

Case 3: $x \in]0, 1[\Rightarrow$ player 1 must be indifferent between T and $B \Rightarrow y \times 6 + (1 - y) \times 2 = y \times 7 \Rightarrow y = \frac{2}{3} \Rightarrow y \in]0, 1[\Rightarrow$ player 2 is indifferent between L and $R \Rightarrow x \times 6 + (1 - x) \times 2 = x \times 7 \Rightarrow x = \frac{2}{3} \Rightarrow x \in]0, 1[$ as it should be. We have found the Nash equilibrium $(\frac{2}{3}T + \frac{1}{3}B, \frac{2}{3}L + \frac{1}{3}R)$.

The second zero-sum game has a unique Nash equilibrium given by $(\frac{7}{10}T + \frac{3}{10}B, \frac{1}{2}L + \frac{1}{2}R)$. It may be found as above, by exploring all possible supports:

Case 1: $x = 1 \Rightarrow y = 0 \Rightarrow x = 0$, a contradiction.

Case 2: $x = 0 \Rightarrow y = 1 \Rightarrow x = 1$, a contradiction.

Case 3: $x \in]0, 1[\Rightarrow$ player 1 must be indifferent between T and $B \Rightarrow y \times 2 + (1 - y) \times (-1) = y \times (-3) + (1 - y) \times 4 \Rightarrow y = \frac{1}{2} \Rightarrow y \in]0, 1[\Rightarrow$ player 2 is indifferent between L and $R \Rightarrow x = \frac{7}{10} \Rightarrow x \in]0, 1[$ as it should be. The unique Nash equilibrium is $(\frac{7}{10}T + \frac{3}{10}B, \frac{1}{2}L + \frac{1}{2}R)$.

The third has two Nash components: (T, R) is a strict and isolated Nash equilibrium, and $\{(xT + (1 - x)B, L)$, with $x \in [0, 1/2]\}$ is a connected component of Nash equilibria.

In the fourth game, after eliminating B and R (strictly dominated), we are left with a 2×2 game with three Nash equilibria (T, L), (M, M) and $(\frac{4}{5}T + \frac{1}{5}M, \frac{4}{5}L + \frac{1}{5}M)$.

(2) The last game has three equilibria: (B, R, W), (T, L, E) and $(\frac{1}{2}T + \frac{1}{2}B, \frac{1}{2}L + \frac{1}{2}R, \frac{1}{2}W + \frac{1}{2}E)$.

Exercise 2

(1) $N = \{1, \ldots, n\}$, $S^i = \mathbb{R}_+$ for all $i \in I$, and $g^i(x^1, \ldots, x^n) = x^i \max(1 - \sum_j x^j, 0)$.

(2) Let $x = (x^1, \ldots, x^n)$ be a Nash equilibrium and, for any $i \in I$, define $s^i := \sum_{j \neq i} x^j$ and $s := \sum_j x^j$. Suppose there is an $i \in I$ such that $s^i < 1$. Then one

shows that $x^i = (1 - s^i)/2$ is a best response for player i against x^{-i}. Consequently, $s < 1$, and so for all $j \in I$, $s^j < 1$ holds, hence $x^j = (1 - s^j)/2$. Consequently, $x^j + s = 1$ for all $j \in I$, implying that $x^1 = \cdots = x^n = \frac{1}{n+1}$. This is a Cournot–Nash equilibrium inducing a price $p = \frac{n}{n+1}$, a total production of $\frac{n}{n+1}$, and a total revenue of $\frac{n}{(n+1)^2}$.

Otherwise, any profile x such that $s^i \geqslant 1$ for all $i \in I$ is a Nash equilibrium inducing a price of zero. (This equilibrium component is unstable: it vanishes as soon as the production cost is positive.)

(3) At the stable equilibrium, the total profit $\frac{n}{(n+1)^2}$ is decreasing with n, the maximum profit $\frac{1}{4}$ is achieved for the monopoly ($n = 1$) and the limiting profit as n increases tends to zero. The equilibrium price $\frac{n}{n+1}$ is decreasing with n and converges to zero (the marginal cost) as n increases.

Exercise 3

(1) In the mixed extension of G, player 1 chooses a probability $(p, 1 - p)$ on $S^1 = \{1, j\}$ that can be one-to-one mapped to the complex number $a = p \times 1 + (1 - p) \times j$. Thus, the set of mixed strategies of player 1 may be viewed as the complex interval $M^1 = [1, j] = \{p1 + (1 - p)j, \ p \in [0, 1]\}$ and similarly, $M^2 = [j, j^2]$ for players 2 and $M^3 = [j^2, 1]$ for player 3. Let $g^1(a, b, c)$ be the expected payoff of player 1 in the game G if the mixed strategy profile is $(a, b, c) \in M^1 \times M^2 \times M^3$. $g^1(a, b, c)$ and $\mathrm{Re}(abc)$ are both multilinear and coincide for pure strategies. Thus, they coincide. Similarly for $g^2(a, b, c)$ and $Im(abc)$ so

$$g^1(a, b, c) = \mathrm{Re}(abc), \quad g^2(a, b, c) = \mathrm{Im}(abc), \quad g^3(a, b, c) = 0.$$

(2) Let $F \subset \mathbb{R}^2$ be the projection on player 1 and 2's coordinates of the set of feasible payoffs. F contains the segments $[1, j], [j, j^2]$ and $[j^2, 1]$ but since $|abc| = |a||b||c| \geqslant 1/8 > 0$ for all $(a, b, c) \in M^1 \times M^2 \times M^3, 0 \notin F$.

Exercise 4

(1.a) Let $D = \{d_1, \ldots, d_n, \ldots\}$ be a denumerable dense set of $\overline{A}$. Define inductively $\alpha_1 = d_1$ and $\alpha_n = \sup\{\alpha_{n-1}, d_n\}$ for all n. One can extract a subsequence $(\alpha_{\phi(n)})_n$ that converges to $\alpha \in S$. Since α is a majorant of D, it is also a majorant of $\overline{D} = \overline{A}$ and so is a majorant of A. Since any majorant β of A is a majorant of D, by induction $\beta \geqslant \alpha_n$ for all n, and thus $\beta \geqslant \alpha$. Consequently, $\alpha = \sup A$.

When $A = S$ one obtains that $\sup S \in S$, thus, S admits a greatest element $\max S$ and similarly, a smallest element $\min S$.

(1.b) Let $A = \{s \in S, f(s) \leqslant s\}$. $A \neq \varnothing$ since S has a greatest element. Define $a = \inf(A) \in S$ (by 1.a). Then, for all $s \in A, a \leqslant s$ and since f is monotonic, $f(a) \leqslant f(s) \leqslant s$. Consequently, $f(a)$ is a minorant of A and so $f(a) \leqslant a$.

By monotonicity of f, $f(f(a)) \leqslant f(a)$, and so $f(a) \in A$. Since a is a minorant of A, $a \leqslant f(a)$, and so $a = f(a)$.

(2.a) Fix a player i and a strategy s_{-i} of the opponents. Then $\mathrm{BR}^i(s^{-i})$ is non-empty and compact, since S^i is compact and g^i upper semi-continuous in s^i. This set is a lattice by supermodularity of g^i.

(2.b) Suppose $s^{-i} \geqslant s'^{-i}$ and consider $t'^i \in BR^i(s'^{-i})$. There exists an x^i in $BR^i(s^{-i})$. We have: $0 \geqslant g^i(x^i \vee t'^i, s^{-i}) - g^i(x^i, s^{-i}) \geqslant g^i(t'^i, s^{-i}) - g^i(x^i \wedge t'^i, s^{-i}) \geqslant g^i(t'^i, s'^{-i}) - g^i(x^i \wedge t'^i, s'^{-i}) \geqslant 0$. Define $t^i = x^i \vee t'^i$.

(2.c) For all $s^{-i} \in S^{-i}$ define $f^i(s^{-i})$ in S^i as the greatest element of $BR^i(s^{-i})$ (see 1.a). Consider the application f from $S = \prod_i S^i$ to S by $f(s) = (f^i(s^{-i}))_i$. f is monotone by 2.b, and so has a fixed point by 2.a. Any fixed point of f is a Nash equilibrium of G.

(3) Put $s^1 = q^1$ and $s^2 = -q^2$. Define a new payoff function as $u^1(s^1, s^2) = s^1 P^1(s^1, -s^2) - C^1(s^1)$, and $u^2(s^1, s^2) = -s^2 P^2(s^1, -s^2) - C^2(-s^2)$. Since $\partial u^1 / \partial s^1$ is monotone in s^2, if $s_1 \geqslant s'_1$ and $s_2 \geqslant s'_2$ then

$$\int_{s'^1}^{s^1} \partial_1 u^1(t^1, s^2) dt^1 \geqslant \int_{s'^1}^{s^1} \partial_1 u^1(t^1, s'^2) dt^1.$$

Consequently, $u^1(s^1, s^2) - u^1(s'^1, s^2) \geqslant u^1(s^1, s'^2) - u^1(s'^1, s'^2)$ and so u^1 has increasing differences in (s_1, s_2). Similarly for u_2 ($\partial u_2 / \partial s_2$ is increasing in s_1). Apply 3.b to conclude.

Exercise 5

The game may be presented as follows:

	A_2	B_2
$A_3: A_1$	(0, 0, 0)	(0, 1, 0)
B_1	(1, 0, 0)	(0, 0, 1)

	A_2	B_2
$B_3: A_1$	(0, 0, 1)	(1, 0, 0)
B_1	(0, 1, 0)	(0, 0, 0)

Pure Nash equilibria are obtained when two of the players choose the same room and the other a different one. This yields six equilibria.

Let x, y and z, respectively, be the probabilities that players 1, 2 and 3 choose the red room and suppose that $x + y + z > 0$.

Let us compute the best response of player 3. By playing A_3 his payoff is $(1 - x)(1 - y)$, and playing B_3 his payoff xy. Since $(1 - x)(1 - y) > xy$ iff $x + y < 1$, if $x + y < 1$, player 3 chooses $z = 1$ and if $x + y > 1$, $z = 0$, and for $x + y = 1$, any z is a best response for player 3. The situation is symmetric for all players.

Suppose that at least two players play completely mixed, for example $x \in]0, 1[$ and $y \in]0, 1[$. Then $y + z = 1$ and $x + z = 1$. Thus, $x = y = 1 - z$, and so $z \in]0, 1[$, so player 3 is indifferent and thus $x + y = 1$, implying that $x = y = z = 1/2$.

The last case is when only one of the players is completely mixing, but then the players who play pure do not choose the same room. We obtain infinitely many equilibria where two players are choosing different rooms and the other is indifferent and is mixing.

Exercise 6

(1) supp $(\tau) \subset$ supp (σ) is clear.
Any $i \in$ supp (σ) minimizes $\sum_j \tau_j \|x_i - y_j\|^2$. But

$$\sum_j \tau_j \|x - y_j\|^2 = \sum_j \tau_j (\|x - z\|^2 + \|z - y_j\|^2 + 2\langle x - z, z - y_j \rangle)$$

$$= \|x - z\|^2 + \sum_j \tau_j \|z - y_j\|^2.$$

Thus supp $(\sigma) \subset \{i, x_i \text{ minimizes } \|x_i - z\|^2\}$.

(2.a) (1) implies supp $(\sigma_N) \subset \{i, x_i \text{ minimizes } \|x_i - x_{N+1}\|^2\}$.

Since $d(x_m, x_{N+1}) \leqslant 2\varepsilon$, $i \in \text{supp}(\sigma_N)$ implies $d(x_i, x_{N+1}) \leqslant 2\varepsilon$ so that $d(x_i, x^*) \leqslant 3\varepsilon$. Similarly for $i \in \text{supp}(\tau_N) \subset \text{supp}(\sigma_N)$.

Since $x_{N+1} = \sum_{i=1}^{N} \tau_N(i) y_i$ and $y_i \in F(x_i)$ for each i, we obtain $x_{N+1} \in \text{co}\{\bigcup_z F(z); z \in B(x^*, 3\varepsilon)\}$.

(2.b) F has a compact graph, hence $F(B(x^*, 3\varepsilon))$ is compact and $K_\varepsilon =_{def} \text{co}\{\bigcup_z F(z); z \in B(x^*, 3\varepsilon)\}$. Hence $x^* \in K_\varepsilon$, for each $\varepsilon > 0$.

Let x in $\bigcap_{\varepsilon > 0} K_\varepsilon$. For each n, there exist (Carathéodory) $z_1^n, \ldots, z_{K+1}^n$ in $B(x^*, 1/n)$ such that $x = \sum_{i=1}^{K+1} \lambda_i^n y_i^n$, with $y_i^n \in F(z_i^n)$ for each i. Extracting convergent subsequences, and since F has a closed graph and $F(x^*)$ is convex, we obtain that $x \in F(x^*)$. So that $x^* \in F(x^*)$.

Exercise 7

(1) Let t be a Nash equilibrium: $G_i(s_i, t_{-i}) \leqslant G_i(t_i, t_{-i})$, for all s and all i. Take the sum in i and deduce that $\Phi(s, t) \leqslant \Phi(t, t)$.

Conversely, assume $\Phi(s, t) \leqslant \Phi(t, t)$, $\forall s \in S$. Take s of the form (s_i, t_{-i}) to deduce that t is an equilibrium.

(2.a) Consider the family $(O_s = \{t \in S; \Phi(s, t) > \Phi(t, t)\})_{s \in S}$. Since $\sum_{i=1}^{n} G_i$ is continuous, $t \mapsto \Phi(t, t)$ is continuous. Since $G_i(s_i, \cdot)$ is continuous on S_{-i} for each s_i, for each i, $t \mapsto \Phi(s, t)$ is continuous. Thus $t \mapsto \Phi(s, t) - \Phi(t, t)$ is continuous and O_s is open.

(2.b) S being compact, we extract a finite subcover $\{O_{s^k}\}_{k \in K}$ satisfying

$$\forall t \in S, \quad \max_{k \in K} \Phi(s^k, t) > \Phi(t, t).$$

(2.c) Θ is well defined since for each t, there exists a k with $\Phi(s^k, t) > \Phi(t, t)$ hence $\sum_l (\Phi(s^l, t) - \Phi(t, t))^+ > 0$. Now the image of t is a convex combination of s^k, hence belongs to S. The continuity of the map $t \mapsto \Phi(s, t) - \Phi(t, t)$ shows that Θ is continuous. Brouwer's theorem implies the existence of a fixed point t^* of Θ.

(2.d) $t^* = \sum_{k \in K} \lambda_k s^k$ with $\lambda_k = \frac{(\Phi(s^k, t^*) - \Phi(t^*, t^*))^+}{\sum_l (\Phi(s^l, t^*) - \Phi(t^*, t^*))^+}$. Thus $\lambda_k > 0$ implies $\Phi(s^k, t^*) > \Phi(t^*, t^*)$. Since $G_i(\cdot, s_{-i})$ is concave on S_i for all s_{-i} and each i, $s \mapsto \Phi(s, t^*)$ is concave in s and thus $\Phi(t^*, t^*) \geqslant \sum_k \lambda_k \Phi(s^k, t^*)$. Hence $\Phi(t^*, t^*) > \Phi(t^*, t^*)$, which gives a contradiction.

9.5 Hints for Chapter 5

Exercise 1

(1) Since G has potential P, the multilinear extension satisfies

$$g^i(\sigma^i, \sigma^{-i}) - g^i(\tau^i, \sigma^{-i}) = P(\sigma^i, \sigma^{-i}) - P(\tau^i, \sigma^{-i}),$$

$$\forall i, \forall \sigma^i, \tau^i \in \Sigma^i, \forall \sigma^{-i} \in \Sigma^{-i},$$

hence the assertion.

(2) Clear. Note that there is a mixed equilibrium.

(4) Consider player i changing from m to n. Then

$$g^i(m, s^{-i}) - g^i(n, s^{-i}) = u^m(t^m(m, s^{-i})) - u^n(t^n(n, s^{-i}))$$

and

$$P(m, s^{-i}) - P(n, s^{-i}) = \sum_k \sum_{r=1}^{t^k(m,s^{-i})} u^k(r) - \sum_k \sum_{r=1}^{t^k(n,s^{-i})} u^k(r).$$

Since $t^m(m, s^{-i}) = 1 + t^m(n, s^{-i})$ and for $k \neq m, n$ we have $t^k(m, s^{-i}) = t^k$ (n, s^{-i}), both quantities are equal.

(5) We check that $\frac{d}{dx^{ik}} W(x) = u^k(z^k)$.

(6) Define $\phi(t) = P(x(t))$. Then

$$\dot{\phi}(t) = \sum_i P(\dot{x}^i(t), x^{-i}(t))$$

$$= \sum_i P(\dot{x}^i(t) + x^i(t) - x^i(t), x^{-i}(t))$$

by linearity and since $\dot{x}^i(t) + x^i(t) \in BR^i[x^{-i}(t)]$, we obtain

$$\dot{\phi}(t) = \sum_i g^i(\dot{x}^i(t) + x^i(t), x^{-i}(t)) - g^i(x^i(t), x^{-i}(t))$$

$$\geqslant 0.$$

Hence ϕ is increasing. Since it is bounded, all accumulation points satisfy $x^* \in$ BR(x^*).

(7) (BNN)

$$\langle \mathcal{B}_\Phi^i(x), \Phi^i(x) \rangle = \sum_{p \in S^i} \left[\widehat{\Phi}_p^i(x) - x_p^i \sum_{q \in S^i} \widehat{\Phi}_q^i(x) \right] \Phi_p^i(x)$$

$$= \sum_{p \in S^i} \widehat{\Phi}_p^i(x) \Phi_p^i(x) - \sum_{p \in S^i} x_p^i \Phi_p^i(x) \sum_{q \in S^i} \widehat{\Phi}_q^i(x)$$

$$= \sum_{p \in S^i} \widehat{\Phi}^i_p(x)\Phi^i_p(x) - \sum_{q \in S^i} \widehat{\Phi}^i_q(x)\overline{\Phi}^i(x)$$

$$= \sum_{p \in S^i} \widehat{\Phi}^i_p(x)[\Phi^i_p(x) - \overline{\Phi}^i(x)]$$

$$= \sum_{p \in S^i} (\widehat{\Phi}^i_p(x))^2 \geq 0.$$

The equality holds if and only if for all $p \in S^i$, $\widehat{\Phi}^i_p(x) = 0$, in which case $\mathcal{B}^i_\Phi(x) = 0$.
(Smith)

$$\langle \mathcal{B}^i_\Phi(x), \Phi^i(x) \rangle = \sum_{p \in S^i} \left(\sum_{q \in S^i} x^i_q [\Phi^i_p(x) - \Phi^i_q(x)]^+ \right) \Phi^i_p(x)$$

$$- \sum_{p \in S^i} x^i_p \Phi^i_p(x) \sum_{q \in S^i} [\Phi^i_q(x) - \Phi^i_p(x)]^+$$

$$= \sum_{p,q} x^i_q \Phi^i_p(x)[\Phi^i_p(x) - \Phi^i_q(x)]^+ - \sum_{q,p} x^i_q \Phi^i_q(x)[\Phi^i_p(x) - \Phi^i_q(x)]^+$$

$$= \sum_{p,q} x^i_q ([\Phi^i_p(x) - \Phi^i_q(x)]^+)^2 \geq 0.$$

The equality holds if and only if for all $q \in S^i$, either $x^i_q = 0$ or $\Phi^i_q(x) \geq \Phi^i_p(x)$ for all $p \in S^i$, in which case $\mathcal{B}^i_\Phi(x) = 0$.
(BR)

$$\langle \mathcal{B}^i_\Phi(x), \Phi^i(x) \rangle = \langle y^i - x^i, \Phi^i(x) \rangle \geq 0$$

since $y^i \in \mathrm{BR}^i(x)$. The equality holds if and only if $x^i \in \mathrm{BR}^i(x)$, hence $\mathcal{B}^i_\Phi(x) = 0$.
Use Proposition 5.5.6 and the fact that the rest points of the dynamics are equilibria.

Exercise 2

(1) Given $x^* \in \mathrm{SNE}(\Phi)$, for all $x \in X$ and $\varepsilon \in]0, 1[$, define $y = \varepsilon x + (1 - \varepsilon)x^*$. Then (5.5) yields

$$\langle \Phi(y), x^* - x \rangle \geq 0, \qquad \forall x \in X,$$

which implies, by continuity,

$$\langle \Phi(x^*), x^* - x \rangle \geq 0, \qquad \forall x \in X.$$

(2) On the other hand, if $x^* \in \mathrm{NE}(\Phi)$ and Φ is dissipative, by adding $\langle \Phi(y) - \Phi(x^*), x^* - y \rangle \geq 0$ and $\langle \Phi(x^*), x^* - y \rangle \geq 0$, we have $x^* \in \mathrm{SNE}(\Phi)$.
(3) Clear from the definition of $\mathrm{SNE}(\Phi)$ (= $\mathrm{NE}(\Phi)$ in this case).

Comments

The same approach can be extended to the case where different types of participants (atomic/non atomic, splittable/non splittable) are present, see Sorin and Wan [204].

This result is proved in Sandholm [184] for population games.

In the framework of population games, Hofbauer and Sandholm [100] introduce this class of games and call them "stable games".

Exercise 3

Denote by x_1, x_2, x_3, y_1, y_2, respectively, the probabilities of T, M, B, L and R.

– If $\alpha < 0$ the strategy T is strictly dominated by $\frac{1}{2}M + \frac{1}{2}B$. The equilibria are those of the zero-sum game:

	L	R
M	1, −1	−1, 1
B	−1, 1	1, −1

This leads to the equilibrium $(x = (0, \frac{1}{2}, \frac{1}{2}), y = (\frac{1}{2}, \frac{1}{2}))$.

– If $\alpha = 0$, we obtain a zero-sum game where both players can guarantee 0. This game has 0 as value and the optimal strategies are $y = (\frac{1}{2}, \frac{1}{2})$ for player 2 and $x = (1 - 2\alpha, \alpha, \alpha)$, for $\alpha \in [0, 1/2]$, for player 1. Any combination of optimal strategies is a Nash equilibrium.

– Suppose now that $\alpha > 0$. If $x_2 > x_3$, because player 2 is playing a best response, $y_2 = 1$ and since player 1 is best replying, $x_2 = 0$, a contradiction. Similarly, $x_3 > x_2$ is impossible at equilibrium, and so, necessarily $x_3 = x_2$. In that case, all strategies against x yield the same payoff to player 2. Since player 1's payoff should be at least $\alpha > 0$, we have $x_2 = x_3 = 0$. Thus $x = (1, 0, 0)$. Since x is a best response against y, $\alpha \geqslant y_1 - (1 - y_1) = 2y_1 - 1$, and $\alpha \geqslant 1 - 2y_1$, which holds if $\frac{(1-\alpha)}{2} \leqslant y_1 \leqslant \frac{(1+\alpha)}{2}$. Thus, if $0 < \alpha < 1$, the Nash equilibria are the profiles $(x = (1, 0, 0), y = (y_1, y_2))$ where $\frac{(1-\alpha)}{2} \leqslant y_1 \leqslant \frac{(1+\alpha)}{2}$ and if $\alpha \geqslant 1$, all $(x = (1, 0, 0), y = (y_1, y_2))$ are Nash equilibria.

Exercise 4

(1) By definition of fictitious play and linearity:

$$\sum_{m \leqslant n-1} F^i(s_n^i, s_m^{-i}) \geqslant \sum_{m \leqslant n-1} F^i(t, s_m^{-i}), \quad \forall t \in S^i.$$

Write $(n - 1)E_n^i = b_n = \sum_{m \leqslant n-1} a(n, m)$ for the left-hand side. By choosing $t = s_{n-1}^i$ we obtain

$$b_n \geqslant a(n - 1, n - 1) + b_{n-1},$$

hence by induction

$$E_n^i \geqslant A_n^i = \sum_{m \leqslant n-1} a(m, m)/(n - 1).$$

(2) If $F^i(s_n^i, s_{n-1}^{-i}) < F^i(s_{n-1})$, by adding to the inequality expressing the fictitious property for stage $n - 1$: $F^i(s_{n-1}^i, \bar{s}_{n-2}^{-i}) \geqslant F^i(s_n^i, \bar{s}_{n-2}^{-i})$, this would contradict the fictitious property for stage n.

(3) Note that the improvement principle implies that the process will stay on Pareto entries. Hence the sum of the stage payoffs will always be $(a + b)$. If fictitious play converges then it converges to $(1/3, 1/3, 1/3)$ so that the anticipated payoff converges to the Nash payoff $\frac{a+b}{3}$, which contradicts the inequality obtained in (1).

9.6 Hints for Chapter 6

Exercise 1

(1) This game is a simple finite perfect information game. By Zermolo's theorem player 1 (the first mover) has a winning strategy or player 2 has a winning strategy. Suppose by contraction that player 2 is winning. If player 1 starts by playing the square (n, m), player 2 has a winning best response $[(k, l); \tau]$. But then, player 1 can imitate player 2's strategy and starts the game by choosing (k, l) followed by τ. Doing so, player 1 is guaranteed to win, a contradiction.

(2) Player 1 starts by playing $(2, 2)$, then each time player 2 chooses a square $(1, k)$ (resp. $(k, 1)$), $k = 2, \ldots, n$, player 1 chooses the symmetric square $(k, 1)$ (resp. $(1, k)$).

(3.a) When $m = \infty$ and $n = 2$, player 2 has a winning strategy. Actually, if player 1 starts by playing $(1, k)$, he loses because the game becomes a $2 \times k$ game where player 2 is the first mover. Otherwise, player 1 starts by playing $(2, k)$, but then player 2 best responds by playing $(1, k + 1)$, and then player 2 makes sure that this configuration repeats itself.

(3.b) When $m = \infty$ and $n \geqslant 3$, player 1 can win by coming back to the $2 \times \infty$ game (by choosing at the first stage the square $(3, 1)$).

Exercise 2

The extensive form is

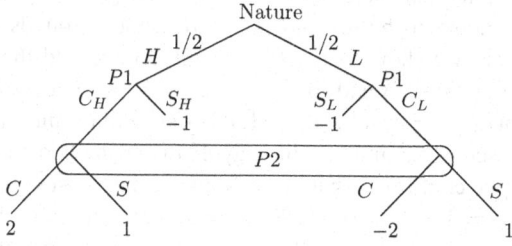

and so, the normal form is

	C	S
(C_H, C_L)	0	1
(C_H, S_L)	1/2	0
(S_H, C_L)	-3/2	0
(S_H, S_L)	-1	-1

The two last rows are strictly dominated (it is never optimal to stop when player 1 has an H card). The remaining two-by-two zero-sum game has a value of 1/3. Player 1 has a unique optimal strategy: $\frac{1}{3}(C_H, C_L) + \frac{2}{3}(C_H, S_L)$, and the corresponding behavioral strategy is: always continue when the card is high, and continue with probability $\frac{1}{3}$ when it is low (bluff $\frac{1}{3}$ of the time). Player 2's optimal strategy is $\frac{2}{3}C + \frac{1}{3}S$ (check the adversary card $\frac{2}{3}$ of the time).

Exercise 3

(1.1) Let σ be fixed and define $\Sigma(\eta)$. Then for all $i \in I$, define $\gamma(\sigma, t^i)$ as the number of pure strategies of player i strictly better that t^i against σ^{-i}. Define τ such that for all i and all s^i,

$$\tau^i(s^i) = \frac{\varepsilon^{\gamma(\sigma, s^i)}}{\sum_{t^i \in S^i} \varepsilon^{\gamma(\sigma, t^i)}}.$$

Then τ is in $F(\sigma)$.

(1.2) By applying Kakutani's fixed point theorem, there is a fixed point σ_ε of F. Then σ_ε is ε-proper and any of its accumulation points as $\varepsilon \to 0$ is a proper equilibrium.

(2) In game 1: (T, L) is proper (and so is perfect and thus is a Nash equilibrium); (B, R) is Nash but not perfect. In game 2, (T, L) is proper and (M, M) is a Nash equilibrium, is perfect, but not proper.

Exercise 4

(1) Player 1 (the proposer) keeps all the money for himself and offers $x = 0$ to player 2. Actually, since player 2 should accept any offer $x > 0$, the unique best response of player 1 that maximizes $M - x$ is $x = 0$.

(2) Let $a_n(s)$ be the subgame-perfect equilibrium payoff of the first proposer when the initial amount to be divided is $s > 0$ in the n rounds alternative offer game. By induction, we show that $a_n(s)$ is well defined and that it is positively homogenous as a function of s and so, $a_n(s) = a_n \times s$ for some positive constant a_n. Part (1) proves that $a_1(s) = s$ and so $a_1 = 1$. Also, at equilibrium $a_2(s) = s - a_1(\delta s)$ because the proposer is maximizing his payoff under the constraint that he must offer to the other player at least what he can obtain if he refuses the proposal. Thus, by homogeneity $a_2 = 1 - \delta a_1$. Similarly, $a_3 = 1 - \delta a_2$ and $a_n = 1 - \delta a_{n-1}$. Thus, $a_n = \sum_{k=0}^{n-1}(-\delta)^k = \frac{1+(-\delta)^n}{1+\delta}$. Hence, if one starts with an amount of money equal to M and the game has T rounds, the proposer gets $\frac{1+(-\delta)^n}{1+\delta}M$, which converges to $\frac{M}{1+\delta}$ as T goes to infinity.

(3) It is easy to check that the proposed profile is a subgame-perfect equilibrium. Let us prove that it is unique. Let $A(s) \subset \mathbb{R}^2$ be the set of subgame-perfect equilibrium

payoffs when player 1 is the first proposer and that the amount of money to be decided is s and let $\overline{a}(s)$ (resp. $\underline{a}(s)$) be the supremum (resp. the infimum) on the first (resp. second) coordinate of $A(s)$. We have $0 \leqslant \underline{a}(s) \leqslant \overline{a}(s) \leqslant s$, and $\overline{a}(s) \geqslant s/(1 + \delta)$. In a subgame-perfect equilibrium, if at the first stage player 1 proposes the sharing $(ts, (1 - t)s)$ for some $t \in [0, 1]$, player 2 accepts if $(1 - t)s > \overline{a}(s\delta)$ or equivalently if $t < 1 - \overline{a}(s\delta)/s$, and player 2 refuses if $(1 - t)s < \underline{a}(s\delta)$ or equivalently if $t > 1 - \underline{a}(s\delta)/s$. Thus player 1 wins at least $s(1 - \overline{a}(s\delta)/s)$, and so

$$\underline{a}(s) \geqslant s - \overline{a}(s\delta).$$

Similarly, in any subgame-perfect equilibrium, player 1 wins at most $s(1 - \underline{a}(s\delta)/s)$ if player 2 accepts the proposition at step 1, and if player 1 refuses, it is even worse for player 1 (who cannot hope for more than $s\delta - \underline{a}(s\delta)$ if he accepts player 2's proposal at step 2). Thus we obtain:

$$\overline{a}(s) \leqslant s - \underline{a}(s\delta).$$

Combining the two inequalities leads to $\overline{a}(s) \leqslant s - (s\delta - \overline{a}(s\delta^2)) \leqslant s(1 - \delta) + s\delta^2(1 - \delta) + \overline{a}(s\delta^4) \leqslant \cdots \leqslant s/(1 + \delta)$. Thus $\overline{a}(s) = s/(1 + \delta)$. Consequently $\underline{a}(s) \geqslant s - (s\delta)/(1 + \delta)$, leading to $\underline{a}(s) = \overline{a}(s) = s/(1 + \delta)$.

Exercise 5

(1) Let σ be a pure strategy equilibrium of G. Let us show that σ is also a pure equilibrium of G'. Since it is pure, it recommends a unique choice in every information set, and because there is no Nature, there is a unique play. To check that σ is a Nash equilibrium of G', it suffices to check that in every information set in G' reached by σ, no player has an interest to deviate. So, let I' be an information set in G' reached by σ. If a player has a profitable deviation in G', it is also profitable in G, a contradiction.

(2) The result is false if σ is mixed. Consider the Matching Pennies game in extensive form. Player 1 starts and chooses between Heads or Tails, and without observing player 1's move player 2 should do the same. If player 2 guesses player 1's choice correctly, player 2 gets $+1$ (and player 1 gets -1). Otherwise player 1 gets $+1$ and player 2 gets -1. The unique equilibrium for both players is mixed: $\frac{1}{2}H + \frac{1}{2}T$. If player 2 gets to know player 1's choice, a random choice is not optimal for him.

Furthermore, the result does not extend with moves of Nature. Just replace in the above game player 1 by Nature, which plays according to the probability distribution $\frac{3}{4}H + \frac{1}{4}T$. Without knowing Nature's choice, player 2 plays H, but knowing Nature's choice, he must play T if he observes T.

Exercise 6

Solve the game using backward induction. If there are 1, 2 or 3 tokens, the player whose turn it is to play wins by taking all the tokens. If 4 tokens remain, the player who starts loses whatever he plays (because after his move, 1, 2 or 3 tokens are left). If there are 5, 6 or 7 tokens, the player who starts can take what is needed so that 4 tokens remain.

Thus, by induction, if the number of tokens is a multiple of 4, the second player has a winning strategy, otherwise the first player will win.

Exercise 7

The game is very similar to the poker game. The extensive form is:

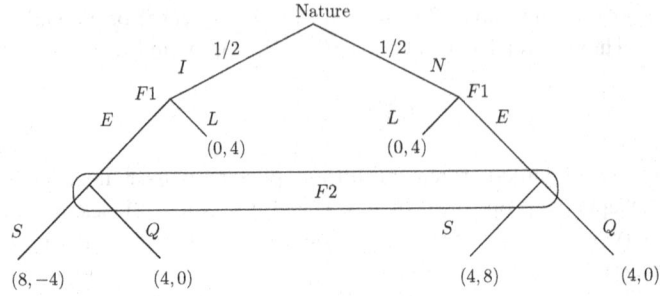

The normal form is:

	E	L
SS	$2, 2$	$4, 0$
SQ	$4, 0$	$2, 2$
QS	$-2, 6$	$2, 2$
QQ	$0, 4$	$0, 4$

QS and QQ are strictly dominated. The remaining games admit a unique equilibrium where each player randomizes equally between the remaining strategies.

Exercise 8

(1) The utility of the buyer is $[v - (b_1 + b_2)/2]\, 1_{b_1 \leqslant b_2}$ and of the seller is $[(b_1 + b_2)/2 - c]\, 1_{b_1 \leqslant b_2}$. The buyer minimizes in b_2 and the seller maximizes in b_1 under the constraint that $b_1 \leqslant b_2$. At equilibrium $b_1 = b_2$, and for all $b \in [c, v]$, $b = b_1 = b_2$ is a Nash equilibrium.

(2) Let us show that $b_i(\cdot)$ is increasing.
Put $\pi_{b_2}(v) = E\left[[v - (b_1(c) + b_2)/2]\, 1_{b_1(c) \leqslant b_2}\right]$.
$\pi_{b_2}(v + \varepsilon) = \pi_{b_2}(v) + \varepsilon P(b_2 \geqslant b_1(c))$.
Define $b_2(v) = \arg \max_{b_2} \pi_{b_2}(v)$.

Then $\pi_{b_2(v)}(v + \varepsilon) = \pi_{b_2(v)}(v) + \varepsilon P(b_2(v) \geqslant b_1(c))$. When $b_2 < b_2(v)$, the first term is smaller because we deviate from the optimum, and the second term is smaller because the bound decreases. The optimum of $b_2 \to \pi_{b_2}(v + \varepsilon)$ is necessarily at the right of $b_2(v)$, that is, $b_2(v + \varepsilon) > b_2(v)$.

(3) Suppose the bidding functions β_1, β_2 are strictly increasing and C^1. Since for each c, $\beta_1(c)$ maximizes the expected payoff

$$\int_{\beta_2^{-1}(\beta_1(c))}^{1} \left(\frac{\beta_1(c) + \beta_2(v)}{2} - c\right) dv,$$

the first-order condition gives:

$$\frac{1}{2}(1 - \beta_2^{-1}(\beta_1(c))) - d\beta_2^{-1}(\beta_1(c))(\beta_1(c) - c) = 0.$$

For the second player, $\beta_2(v)$ maximizes the expected payoff:

$$\int_0^{\beta_1^{-1}(\beta_2(v))} \left(v - \frac{\beta_1(c) + \beta_2(v)}{2}\right) dc.$$

The first-order condition gives:

$$-\frac{1}{2}\beta_1^{-1}(\beta_2(v)) + d\beta_1^{-1}(\beta_2(v))(v - \beta_2(v)) = 0.$$

This gives a pair of ODEs.

(4) If we look for affine solutions we find:

$$\beta_1(c) = \frac{2}{3}c + \frac{1}{4}$$

and

$$\beta_2(v) = \frac{2}{3}v + \frac{1}{12}.$$

There is a trade if and only if $v - c \geqslant 1/4$.

9.7 Hints for Chapter 7

Exercise 1

(1) The normal form is:

	(B_B, B_W)	(B_B, W_W)	(W_B, B_W)	(W_B, W_W)
B	(2, 2)	(2, 2)	(5/2, 5/2)	(5/2, 5/2)
W	(5/2, 5/2)	(2, 2)	(5/2, 5/2)	(2, 2)

There are infinitely many Nash equilibria forming one single component, and they all give to both players a payoff of $\frac{5}{2}$. In the extensive form, all equilibria are such that if player 1 chooses one of the colors, player 2 (who observes player 1's choice) plays the opposite color.

(2) The normal form is:

	(B_B, B_W)	(B_B, W_W)	(W_B, B_W)	(W_B, W_W)
(B_B, B_W)	$(2, 2)$	$(2, 2)$	$(5/2, 5/2)$	$(5/2, 5/2)$
(B_B, W_W)	$(7/2, 1)$	$(2, 2)$	$(5, 0)$	$(7/2, 1)$
(W_B, B_W)	$(1, 7/2)$	$(2, 2)$	$(0, 5)$	$(1, 7/2)$
(W_B, W_W)	$(5/2, 5/2)$	$(2, 2)$	$(5/2, 5/2)$	$(2, 2)$

In any pure or mixed equilibrium, player 2 plays (B_B, W_W). The unique equilibrium payoff is $(2, 2)$.

(3) When both players are informed, the unique equilibrium payoff is also $(2, 2)$. When only player 2 is informed the unique equilibrium payoff is $(1, 7/2)$. The best for player 1 is game 1 where no player is informed, in which case he gets 1. The value of information is negative because if we only inform player 1, his payoff decreases.

Exercise 2

(2) The normal form is:

	$\begin{matrix}L_A\\L_B\end{matrix}$	$\begin{matrix}L_A\\M_B\end{matrix}$	$\begin{matrix}L_A\\R_B\end{matrix}$	$\begin{matrix}M_A\\L_B\end{matrix}$	$\begin{matrix}M_A\\M_B\end{matrix}$	$\begin{matrix}M_A\\R_B\end{matrix}$	$\begin{matrix}R_A\\L_B\end{matrix}$	$\begin{matrix}R_A\\M_B\end{matrix}$	$\begin{matrix}R_A\\R_B\end{matrix}$
$\begin{matrix}A_1\\A_2\end{matrix}$	$(0, 3)$	$(0, 3)$	$(0, 3)$	$(2, 5)$	$(2, 5)$	$(2, 5)$	$(1, 6)_*$	$(1, 6)$	$(1, 6)_*$
$\begin{matrix}A_1\\B_2\end{matrix}$	$(0, 3)$	$(1, 11/2)$	$(1, 9)_*$	$(1, 5/2)$	$(2, 5)$	$(2, 17/2)$	$(0, 0)$	$(1, 5/2)$	$(1, 6)$
$\begin{matrix}B_1\\A_2\end{matrix}$	$(0, 3)$	$(1, 5/2)$	$(0, 0)$	$(2, 5)$	$(2, 5)$	$(2, 17/2)$	$(1, 9)_*$	$(2, 17/2)$	$(1, 6)$
$\begin{matrix}B_1\\B_2\end{matrix}$	$(0, 3)$	$(2, 5)$	$(1, 6)_*$	$(0, 3)$	$(2, 5)$	$(1, 6)$	$(0, 3)$	$(2, 5)$	$(1, 6)_*$

Pure Nash equilibria are indicated with a $*$. They all give to player 1 a payoff of 1. The pure equilibria (A_1, A_2) and (B_1, B_2) are called pooling or non-revealing because whatever information he has, player 1 plays the same action, and so after observing player 1's action, player 2's probabilistic belief about the state of Nature remains the same. The pure equilibria (A_1, B_2) and (B_1, A_2) are called completely revealing because player 1 plays a different action after every state, and so after observing player 1's action, player 2 knows the state with certainty.

(3) The suggested behavioral strategy profile corresponds to the mixed equilibrium $(\frac{1}{2}(A_1, A_2) + \frac{1}{2}(A_1, B_2), (M_A, R_B))$. The corresponding vector payoff is $(2, 27/4)$ and so player 1's payoff improves. In this equilibrium, player 2 learns the state if he observes action B and after observing action A his belief changes from the probability $(\frac{1}{2}, \frac{1}{2})$ on the state to $(\frac{2}{3}, \frac{1}{3})$.

Remark: In a general "sender-receiver" game in which player 1, the sender, learns the state of Nature, then sends a non-costly message to player 2, who then takes an action, there is always a non-revealing equilibrium. A completely revealing equilibrium, if it exists, gives the best possible payoff to player 2. The best equilibrium for player 1 may be partially revealing, as in this example.

Exercise 3

(1) By contradiction let $((x_1, x_2, x_3), (y_1, y_2, y_3))$ be an equilibrium. If $y_1 > 0$, then $x_1 = 0$ so that $y_3 = 0$. If moreover $y_2 > 0$, then $x_2 = 1$ and then y_1, a contradiction. Thus $y_1 = 0$, which implies $x_3 = 0$. If $x_2 > 0$, $y_2 = 1$ and then $x_1 = 1$. If $x_2 = 0$, then $x_1 = 1$, also leading to a contradiction.

(2.a) There is no pure equilibria (what would $\sum_i s^i$ be?).

(2.b) By contradiction, consider an equilibrium inducing the *independent* random variables $s^1, s^2, \ldots, s^n, \ldots$ By the Borel–Cantelli lemma, the probability that $\sum_i s^i = \infty$ is 0 or 1. Both cases are impossible.

(2.c) Consider player i and let s^i be his signal. We have $P(\mu_1|s^i = 1) = P(\mu_2|s^i = 1) = P(\mu_1|s^i = 0) = P(\mu_2|s^i = 0) = 1/2$, hence player i is indifferent and always plays a best response.

Exercise 4

(1) Clear since Γ is a finite zero-sum game.

(2) If $Q \in \Delta(S)$ is a strategy of player 1 that guarantees 0 in Γ, then Q is a correlated equilibrium distribution in G.

(3) We have $\gamma(\mu, \pi)$ is equal to

$$\sum_{(s^1,s^2) \in S^1 \times S^2} \mu^1(s^1)\mu^2(s^2) \sum_{i=1,2} \sum_{(t^i,u^i) \in L^i} \pi(t^i, u^i)\gamma((s^1, s^2); t^i, u^i).$$

Let A_1 be the term corresponding to $i = 1$. Then

$$
\begin{aligned}
A_1 &= \sum_{s^1,s^2} \mu^1(s^1)\mu^2(s^2) \sum_{(t^1,u^1)} \pi(t^1, u^1)\gamma((s^1, s^2); t^1, u^1) \\
&= \sum_{s^1} \mu^1(s^1) \sum_{u^1} \pi(s^1, u^1) \sum_{s^2} \mu^2(s^2)\gamma((s^1, s^2); s^1, u^1) \\
&= \sum_{s^1} \mu^1(s^1) \sum_{u^1 \neq s^1} \rho^1(s^1, u^1) \sum_{s^2} \mu^2(s^2)(g^1(s^1, s^2) - g^1(u^1, s^2)) \\
&= \sum_{s^1} \mu^1(s^1) \sum_{u^1} \rho^1(s^1, u^1)(g^1(s^1, \mu^2) - g^1(u^1, \mu^2)) \\
&= \sum_{s^1,u^1} \mu^1(s^1)\rho^1(s^1, u^1)g^1(s^1, \mu^2) - \sum_{s^1,u^1} \mu^1(s^1)\rho^1(s^1, u^1)g^1(u^1, \mu^2) \\
&= \sum_{s^1} \mu^1(s^1)g^1(s^1, \mu^2) - \sum_{u_1} \mu^1(u^1)g^1(u^1, \mu^2) \\
&= g^1(\mu^1, \mu^2) - g^1(\mu^1, \mu^2) = 0.
\end{aligned}
$$

Similarly $A_2 = 0$, hence $\gamma(\mu, \pi) = 0$.

(4) $\forall \pi \in \Delta(L), \exists \mu \in \Delta(S)$ such that $\gamma(\mu, \pi) \geqslant 0$, hence the value of Γ is nonnegative, thus (a) and (b) apply.

(5) In the auxiliary two-person zero-sum game Γ, the strategy space of player 1 is $S = S^1 \times S^2 \times \cdots \times S^n$, that of player 2 is $\bigcup_{i=1}^n (S^i \times S^i)$ and the payoff is defined in a similar way.

Exercise 5

(1.a) Let $\pi \in \text{CED}$. We have, for player 1:

$$
\begin{aligned}
g(\pi) &= \sum_{s^1 \in S^1} \sum_{s^2 \in S^2} g(s^1, s^2) \pi(s^1, s^2) \\
&= \sum_{s^1 \in S^1} \max_{t^1 \in S^1} \sum_{s^2 \in S^2} g(t^1, s^2) \pi(s^1, s^2) \\
&= \sum_{s^1 \in S^1} \pi(s^1) \max_{t^1 \in S^1} G^1(t^1, \pi(\cdot|s^1)).
\end{aligned}
$$

But $\max_{t^1 \in S^1} g(t^1, \pi(\cdot|s^1)) \geqslant v$ for all s^1, thus $g(\pi) \geqslant v$. By symmetry $g(\pi) = v$.

(1.b) Let $\pi \in \text{CED}$ and s^1 be an action of player 1 having a positive probability under π. We have $v = g(\pi) = \sum_{s^1 \in S^1} \pi(s^1) \max_{t^1 \in S^1} g(t^1, \pi(\cdot|s^1))$, with $\max_{t^1 \in S^1} g(t^1, \pi(\cdot|s^1)) \geqslant v$ for all s^1. Hence $\pi(s^1) > 0$ implies

$$
\max_{t^1 \in S^1} G^1(t^1, \pi(\cdot|s^1)) = v,
$$

which means that $\pi(\cdot|s^1)$ is an optimal strategy of player 2 in g.

(2.a) If player 1 receives the signal a_1, his conditional probability on S^2 is $(1/2, 1/2, 0)$ and a_1 is a best response. If the signal is a_2, his conditional probability on S^2 is $(1, 0, 0)$ and a_2 is a best response. The distribution belongs to CED.

(2.b) The value is 0 and the optimal strategies are: $\alpha a_1 + (1 - \alpha)a_2$, with $\alpha \geqslant 1/2$ for player 1 and symmetrically for player 2.

Let $\pi \in \text{CED}$. For each action a of player 1, with $\pi(a) > 0$, (1.b) implies $\pi(b_3|a) = 0$ and $\pi(b_1|a) \geqslant \pi(b_2|a)$.

Thus $\pi(b_3) = 0$, hence also $\pi(a_3) = 0$. π is thus of the form

	b_1	b_2	b_3
a_1	x	y	0
a_2	z	t	0
a_3	0	0	0

with $x \geqslant y$, $z \geqslant t$, $x \geqslant z$ and $y \geqslant t$.

Conversely, we easily check that any distribution as above is in CED.

Exercise 6

(1) The equilibria are $\alpha = (T, L)$, $\beta = (B, R)$ and $\gamma = (\frac{1}{2}T + \frac{1}{2}B, \frac{1}{2}L + \frac{1}{2}R)$.

(2.a) The payoffs of player 1 and 2 are independent of the action of player 3, hence they play an equilibrium of the previous two-person game. If they play α or β, player 3 has no best response. If they play γ, $z = 0$ is not a best response either.

(2.b) The distribution defined by $z = 0$ and the correlated equilibria for players 1 and 2: $\frac{1}{2}$ on (T, L) and $\frac{1}{2}$ on (B, R), gives a payoff 2 to player 3. By playing $z \neq 0$ player 3 obtains at most $\frac{3}{2}$.

The distribution is thus in CED.

9.8 Hints for Chapter 8

Exercise 1
(1) $v^1 = v^2 = 0$, hence $E = \text{co}(g(A)) = \text{co}\{(0, 0), (1, 0), (0, 1)\}$.
(2) $v^1 = 1/2$, $v^2 = 1$, hence $E = \{(x, 1), 1/2 \leqslant x \leqslant 1\}$.
(3) $E = \{(0, 0)\}$.
(4) $v^1 = v^2 = v^3 = 0$, hence $E = \text{co}(g(A)) = \{(x, y, z) \in \mathbb{R}^3_+, x + y + z \leqslant 1\}$.

Exercise 2
Let $v = v^1 = -v^2$ denote the value of G in mixed strategies. Any feasible and IR payoff (x, y) satisfies $x + y = 0$, $x \geqslant v$ and $y \geqslant -v$, so there is a unique feasible and IR payoff, which is $(v, -v)$, and $E_T = E_\lambda = E_\infty = \{(v, -v)\}$ for each T and λ.

Exercise 3
(a) Consider any bounded sequence $(x_t)_{t \geqslant 1}$ of real numbers, and let $\bar{x}_t$ be $\frac{1}{t} \sum_{s=1}^{t} x_s$ for each $t \geqslant 1$. Recall that by (8.1), we have for each $\lambda \in (0, 1]$:

$$\sum_{t=1}^{\infty} \lambda^2 t (1 - \lambda)^{t-1} \bar{x}_t = \lambda \sum_{t=1}^{\infty} (1 - \lambda)^{t-1} x_t.$$

Moreover, for each $T_0 \geqslant 1$,

$$\sum_{t=1}^{T_0-1} \lambda^2 t (1 - \lambda)^{t-1} = -\lambda^2 \frac{\partial}{\partial \lambda} \left(\sum_{t=1}^{T_0-1} (1 - \lambda)^t \right),$$

$$\sum_{t=1}^{T_0-1} \lambda^2 t (1 - \lambda)^{t-1} = 1 - (1 - \lambda)^{T_0} - T_0 \lambda (1 - \lambda)^{T_0-1} \xrightarrow{\lambda \to 0} 0. \qquad (9.2)$$

(b) Let now σ be a uniform equilibrium of G_∞, and fix i in N. For each $t \geqslant 1$, define $x_t^i = \mathbb{E}_\sigma(g^i(a_t))$, so that $\gamma_t^i(\sigma) = \bar{x}_t$.

Consider τ^i in Σ^i and write $y_t = \mathbb{E}_{\tau^i, \sigma^{-i}}(g^i(a_t))$. Fix $\varepsilon > 0$. There exists a T_0 (independent of i and τ^i) such that $\bar{y}_t^i \leqslant \bar{x}_t + \varepsilon$ for each $t \geqslant T_0$. We have for each λ:

$$\gamma_\lambda^i(\tau^i, \sigma^{-i}) = \lambda \sum_{t=1}^\infty (1-\lambda)^{t-1} y_t = \sum_{t=1}^\infty \lambda^2 t (1-\lambda)^{t-1} \overline{y}_t$$

$$\leqslant \sum_{t=1}^{T_0-1} \lambda^2 t (1-\lambda)^{t-1} \overline{y}_t + \sum_{t=T_0}^\infty \lambda^2 t (1-\lambda)^{t-1} (\overline{x}_t + \varepsilon)$$

$$\leqslant \gamma_\lambda^i(\sigma) + \varepsilon + \sum_{t=1}^{T_0-1} \lambda^2 t (1-\lambda)^{t-1} (\overline{y}_t - \overline{x}_t)$$

$$\leqslant \gamma_\lambda^i(\sigma) + \varepsilon + 2\|g^i\|_\infty (1 - (1-\lambda)^{T_0} - T_0\lambda(1-\lambda)^{T_0-1})$$

$$\leqslant \gamma_\lambda^i(\sigma) + 2\varepsilon,$$

where the second equality uses (8.1), and the last inequalities use (9.2) and hold for $\lambda \leqslant \lambda_0$, for some λ_0 depending on ε and T_0 but independent of i and τ^i. This proves (1′) of Lemma 8.3.5.

Finally, (8.1) and (9.2) also imply that $\gamma_\lambda^i(\sigma) = \lambda \sum_{t=1}^\infty (1-\lambda)^{t-1} x_t$ converges, as λ goes to 0, to $\lim_{t\to\infty} \overline{x}_t$. This proves (2′) of the lemma.

Exercise 4

Since we have two players,

$$v^1 = \max_{x^1 \in \Delta(A^1)} \min_{x^2 \in \Delta(A^2)} g^1(x^1, x^2)$$

and similarly $v^2 = \max_{x^2 \in \Delta(A^2)} \min_{x^1 \in \Delta(A^1)} g^2(x^1, x^2)$. The proof is close to the proof in Example 8.2.3, by induction on T. Assume $E_T = \{(v^1, v^2)\}$ and consider a Nash equilibrium $\sigma = (\sigma^1, \sigma^2)$ of G_{T+1}. Let x^1 and x^2 be the mixed actions used at stage 1, respectively under σ^1 and σ^2. The payoff induced by σ in the $T+1$-stage game is $\frac{1}{T+1}(g^1(x^1, x^2) + Tv^1)$. If player 1 deviates and plays a mixed action y^1 at stage 1 and then independently at each subsequent stage an action achieving the maximum in the expression of v^1, he obtains a payoff not lower than $\frac{1}{T+1}(g^1(y^1, x^2) + Tv^1)$. Consequently, $g^1(y^1, x^2) \leqslant g^1(x^1, x^2)$ and x^1 is a best response to x^2 in G. Similarly x^2 is a best response to x^1 in G, and $g(x^1, x^2) = (v^1, v^2)$.

Exercise 5

	L
T	$(0, 0)$
B	$(1, 1)$

Exercise 6

(1) The set of mixed Nash equilibrium payoffs is the union of the segments $[(0, 0), (0, 2)]$ and $[(0, 0), (2, 0)]$, and the set of correlated equilibrium payoffs is the convex hull of the latter.

(2) We need (B, L) and (T, R) to be played with probability 1/2 each at stage 2. Stage 1 can be used by the players to determine whether (B, L) or (T, R) will be played at stage 2, the difficulty being that player 1 strictly prefers (B, L) whereas player 2 strictly prefers (T, R). This difficulty can be solved using a *jointly controlled lottery* (introduced by Aumann and Maschler in the 1960s [13]): at stage 1, both players randomize with probability 1/2 for each action. At stage 2, if the outcome of stage 1 was on the first diagonal, i.e. either (T, L) or (B, R), player 1 plays B and player 2 plays L, and if the outcome of stage 1 was either (B, L) or (T, R), player 1 plays T and player 2 plays R. No player can unilaterally influence the fact that each diagonal will be reached with probability 1/2 at stage 1, and we have defined a Nash equilibrium of Γ with payoff $(1, 1)$.

Exercise 7

Fix any strategy of player 1, and consider a best response of player 2 in G_λ. After any history reached with positive probability, player 2 should play D^2 with probability one, because it maximizes the current stage payoff and does not harm the best response continuation payoff of player 2. Consequently any equilibrium payoff is in the convex hull of $\{(0, 4), (1, 1)\}$. But any equilibrium payoff should be individually rational for player 1, so $E_\lambda = \{(1, 1)\}$.

Exercise 8

The mixed Nash equilibrium payoffs of the stage game are $(2, 1)$, $(1, 2)$ and $(2/3, 2/3)$. In the repeated game, one can use the different stages to convexify these payoffs, and we obtain that $\text{co}\{(2, 1), (1, 2), (2/3, 2/3)\}$ is a subset of $\bigcup_{T \geqslant 1} E_T$. Conversely, fix T and consider a Nash equilibrium $\sigma = (\sigma^1, \sigma^2)$ of G_T. After any history of length t of player 1 having positive probability, σ^1 has to play at stage $t + 1$ a best response in the stage game against the expected mixed action of stage $t + 1$ of player 2. Similarly, after any history of length t of player 2 having positive probability, σ^2 plays at stage $t + 1$ a best response in the stage game against the expected mixed action of stage $t + 1$ of player 1. The sets of best replies being convex, it follows that the expected mixed action profile of stage $t + 1$ is a mixed Nash equilibrium of the stage game, and the payoff of σ is a convex combination of mixed Nash equilibrium payoffs of G. We obtain $\bigcup_{T \geqslant 1} E_T = \text{co}\{(2, 1), (1, 2), (2/3, 2/3)\}$.

Remark: The above arguments are not specific to the battle of the sexes, and for any repeated game played in the dark, $\bigcup_{T \geqslant 1} E_T$ is the convex hull of the mixed Nash equilibrium payoffs of the stage game.

Exercise 9

(1) All minmax are 0, so the set of feasible and IR payoffs is the square given by the convex hull of $(0, 0, 0)$, $(1, 0, 0)$, $(0, 1, 0)$ and $(1, 1, 0)$.

(2) Fix T and a Nash equilibrium $\sigma = (\sigma^1, \sigma^2, \sigma^3)$ of the T-stage repeated game with payoff $(x^1, x^2, 0)$. Define τ^1 as the deviation of player 1, who plays B at every stage, and similarly let τ^2 be the deviation of player 2, who plays R at every stage. The point is that both $(\tau^1, \sigma^2, \sigma^3)$ and $(\sigma^1, \tau^2, \sigma^3)$ induce a certain signal s' for player 3 at every stage. Let $A^3 = \{W, M, E\}$ be the set of pure actions of player 3,

and define $z \in \Delta(A^3)$ as the expected frequency played by player 3 under $(\tau^1, \sigma^2, \sigma^3)$ and $(\sigma^1, \tau^2, \sigma^3)$:

$$\forall a \in A_3, z(a) = \mathbb{E}_{(\tau^1, \sigma^2, \sigma^3)} \left(\frac{1}{T} \sum_{t=1}^{T} \mathbf{1}_{a_t^3 = a} \right) = \mathbb{E}_{(\tau^1, \sigma^2, \sigma^3)} \left(\frac{1}{T} \sum_{t=1}^{T} \mathbf{1}_{a_t^3 = a} \right).$$

We have

$$\mathbb{E}_{(\tau^1, \sigma^2, \sigma^3)} \left(\frac{1}{T} \sum_{t=1}^{T} g^1(a_t) \right) = z(W) + z(E),$$

$$\mathbb{E}_{(\sigma^1, \tau^2, \sigma^3)} \left(\frac{1}{T} \sum_{t=1}^{T} g^2(a_t) \right) = z(W) + z(M).$$

By the best response conditions, $x^1 \geqslant z(W) + z(E)$ and $x^2 \geqslant z(W) + z(M)$, so $x^1 + x^2 \geqslant 1$.

(3) All vectors $(1, 0, 0)$, $(0, 1, 0)$ and $(1, 1, 0)$ are Nash equilibrium payoffs of the stage game. One can use the stages to convexify the equilibrium payoff set, so given $\varepsilon > 0$ there exists a T such that each point of the triangle $\text{co}\{(1, 0, 0), (0, 1, 0), (1, 1, 0)\}$ is at distance at most ε to E_T. Using (a) and (b), we get the convergence of E_T to the triangle.

Remark: Here the limit equilibrium payoff set depends on the "joint rationality" condition: $x^1 + x^2 \geqslant 1$. This condition comes from the fact that in case of deviation, player 3 cannot deduce from his signals who among player 1 or player 2 deviated, and is unable to punish *at the same time* players 1 and 2. So in the case of a deviation, he does not know which player he should punish, and if he punishes player 1 (resp. player 2), he rewards player 2 (resp. player 1). A variant of this example can be found in Tomala [211] (see also [169]). A general characterization of communication equilibrium payoffs of repeated games with signals, based in particular on joint rationality, can be found in Renault and Tomala [170].

Exercise 10

(1) The "only if" part is clear. For the "if" part, assume by contradiction that σ is not a subgame-perfect equilibrium of G_λ. There exists a history h, a player i and a strategy τ^i such that

$$\gamma_\lambda^i(\sigma^{-i}[h], \tau^i[h]) > \gamma_\lambda^i(\sigma[h]).$$

Since $\lambda > 0$ is fixed, the very large stages play only a small rôle and we can assume w.l.o.g. that τ^i differs from σ^i at finitely many information sets only. This allows us to define l^* as the smallest positive integer l such that there exists a strategy $\widehat{\tau}^i$ of player i, different from σ^i after exactly l histories, and satisfying

$$\gamma_\lambda^i(\sigma^{-i}[h], \widehat{\tau}^i[h]) > \gamma_\lambda^i(\sigma[h]).$$

Let τ^{*i} be different from σ^i after exactly l^* histories (denoted $h_1, h_2, \ldots, h_{l^*}$), and such that $\gamma_\lambda^i(\sigma^{-i}[h], \tau^{*i}[h]) > \gamma_\lambda^i(\sigma[h])$. By minimality of l^* each history $h_1, h_2, \ldots, h_{l^*}$ extends h, and we consider a history h^* with maximum length among $h_1, h_2, \ldots, h_{l^*}$.

We have $\gamma_\lambda^i(\sigma^{-i}[h^*], \tau^{*i}[h^*]) > \gamma_\lambda^i(\sigma[h^*])$, otherwise modifying τ^{*i} at h^* only by $\tau^{*i}(h^*) = \sigma^i(h^*)$ would contradict the minimality of l^*. Now, since h^* has maximum length $\tau^{*i}[h^*]$ and $\sigma^i[h^*]$ differ only immediately after h^*. Finally, define τ^{**i} as the strategy of player i, who plays like σ^i everywhere except immediately after h^*, where they play $\tau^{*i}(h^*)$. τ^{**i} is a one-shot deviation of σ^i, and $\gamma_\lambda^i(\sigma^{-i}[h^*], \tau^{**i}[h^*]) > \gamma_\lambda^i(\sigma[h^*])$.

(2.a) Let x in $\mathbb{R}^N$ be in NEP$(\lambda g + (1 - \lambda)u)$, for some $u : A \to E_\lambda'$. One can construct a subgame-perfect equilibrium of G_λ as follows: play at stage 1 a mixed Nash equilibrium in NE$(\lambda g + (1 - \lambda)u)$ with payoff x, and from stage 2 on a subgame-perfect equilibrium of G_λ with payoff $u(a_1)$, where a_1 is the action profile played at stage 1. This shows that $\Phi(E_\lambda') \subset E_\lambda'$.

Conversely, consider a subgame-perfect equilibrium σ of G_λ with payoff w. Denote by x the mixed action profile played at stage 1 under σ. We have $w = \sum_{a \in A} x(a)(\lambda g(a) + (1 - \lambda)\gamma_\lambda(\sigma[a]))$. Define for each a in A, $u(a) = \gamma_\lambda(\sigma[a]) \in E_\lambda'$. x belongs to NE$(\lambda g + (1 - \lambda)u)$, so $w \in$ NEP$(\lambda g + (1 - \lambda)u)$ belongs to $\Phi(E_\lambda')$.

(2.b) Let W be a fixed point of Φ, and fix some u_0 in W. There exists a mixed action profile x_0 and $u_1 : A \to W$ such that x_0 is in NE$(\lambda g + (1 - \lambda)u_1)$ with payoff u_0. For each a in A, $u_1(a) \in W$, so there exists a mixed action profile $x_1(a)$ and $u_2(a) : A \to W$ such that $x_1(a)$ is in NE$(\lambda g + (1 - \lambda)u_2(a))$ with payoff $u_1(a)$.

By induction, for each history h in H_T, there is a mixed action profile $x_T(h)$ and $u_{T+1}(h) : A \to W$ such that $x_T(h)$ is in NE$(\lambda g + (1 - \lambda)u_{T+1}(h))$ with payoff $u_T(h)$, with $u_{T+1}(h, a) := u_{T+1}(h)(a)$ for all T and $h \in H_T$ and $a \in A$.

Let σ be the strategy profile where each player i plays after any history h of length T the mixed action $x_T^i(h)$. By construction, there is no one-shot profitable deviation from σ. By the one-shot deviation principle of (a), σ is a subgame-perfect equilibrium of G_λ, and its payoff is w.

Remark: $W \subset \Phi(W)$ is enough to deduce that $W \subset E_\lambda'$.

Exercise 11

Proof of Lemma 8.5.7.

We use $\| \cdot \| = \| \cdot \|_\infty$, and let N denote the number of players. Fix $\varepsilon > 0$ and a player i in N. There exists an $L \geqslant 1$ such that each point in $\operatorname{co} g(A)$ is $\varepsilon/2$-close to a point in the grid $D^L = \{\frac{1}{L}(g(a_1) + \cdots + g(a_L)), a_1, \ldots, a_L \in A\}$, and there exists a $\lambda_0 > 0$ such that for each $\lambda \leqslant \lambda_0$, each point in $\operatorname{co} g(A)$ is ε-close to a point in the grid:

$$D_\lambda^L = \{\frac{\lambda}{1 - (1 - \lambda)^L} \sum_{t=1}^L (1 - \lambda)^{t-1} g(a_t), a_1, \ldots, a_L \in A \text{ s.t.}$$

$$g^i(a_1) \leqslant g^i(a_2) \leqslant \cdots \leqslant g^i(a_L)\}.$$

Consider λ small enough, and define $\mu = 1 - (1 - \lambda)^L$. There exists a positive integer K such that β, defined as $1 - (1 - \mu)^K$, is in $[1/2, 2/3]$. Consider $d = (d^j)_{j \in N}$ in $\mathbb{R}^N$ such that $B_\infty(d, (3N + 8)\varepsilon) \subset \text{cog}(A)$. Define d^- with i-coordinate $d^i - (N + 3)\varepsilon$, and j-coordinate d^j for each $j \neq i$, and let c be a point in D_λ^L such that $\|c - d^-\| \leqslant \varepsilon$. We have $c^i \leqslant d^i - (N + 2)\varepsilon$, and $\|c - d\| \leqslant (N + 4)\varepsilon$. Define now by f the convex combination $d = \beta c + (1 - \beta)f$. $\|d - f\| \leqslant 2(N + 4)\varepsilon$, so $B(f, N\varepsilon) \subset \text{cog}(A)$. Notice that $f^i \geqslant d^i + (N + 2)\varepsilon$.

We first show that $B(f, N\varepsilon) \subset \text{cog}(A)$ implies that f can be written as a convex combination of points in D_λ^L which are all $(N + 1)\varepsilon$ close to f. Indeed, define for each j in N, $f^+(j)$ and $f^-(j)$ by respectively adding and subtracting $N\varepsilon$ to/from the j-coordinate of f. Clearly, f is in the convex hull of $\{f^+(j), j \in N\} \cup \{f^-(j), j \in N\}$, and we now slightly perturb this convex hull. Let $e^+(j)$ and $e^-(j)$ be points on the grid D_λ^L which are ε-close to, respectively, $f^+(j)$ and $f^-(j)$. Then f belongs to the convex hull of the set $S = \{e^+(j), j \in N\} \cup \{e^-(j), j \in N\}$ (if not, separate f from this convex hull to get the existence of a y in $\mathbb{R}^N$ such that $\langle y, f - e \rangle < 0$ for each e in S, then use the Cauchy–Schwarz inequality and $\|\cdot\|_2 \leqslant \sqrt{N}\|\cdot\|_\infty$ in $\mathbb{R}^N$ to get a contradiction). Now, by Carathéodory's theorem, there exists a subset $S' = \{e_1, \ldots, e_{N+1}\}$ of $N + 1$ elements of S such that f can be written as a convex combination:

$$f = \sum_{m=1}^{N+1} \alpha_m e_m.$$

For λ small enough, we have $\mu \leqslant \frac{1}{N+1}$ so there exists an m_1 such that $\alpha_{m_1} \geqslant \mu$. We can write $f = \mu e(1) + (1 - \mu)f(1)$, with $e(1) = e_{m_1}$ and $f(1)$ in $\text{co}(S')$. Similarly, there exists $e(2)$ in S' and $f(2)$ in $\text{co}(S')$ such that $f(1) = \mu e(2) + (1 - \mu)f(2)$. Iterating, we obtain a sequence $(e(s))_{s \geqslant 1}$ of points in S' such that

$$f = \sum_{s=1}^{\infty} \mu(1 - \mu)^{s-1}e(s).$$

Since $d = \beta c + (1 - \beta)f$, we have

$$d = (1 - (1 - \mu)^K)c + (1 - \mu)^K \sum_{s=1}^{\infty} \mu(1 - \mu)^{s-1}e(s),$$

with

$$\|c - d\| \leqslant (N + 4)\varepsilon, \|e(s) - d\| \leqslant 3(N + 3)\varepsilon, c^i \leqslant d^i - (N + 2)\varepsilon \text{ and}$$
$$e^i(s) \geqslant f^i - \|f - e(s)\| \geqslant d^i + \varepsilon$$

for each s.

By definition of D_λ^L, we can write $c = \frac{\lambda}{1-(1-\lambda)^L} \sum_{t=1}^{L} (1-\lambda)^{t-1} g(a_t)$, with $a_1, \ldots, a_L \in A$ and $g^i(a_1) \leqslant g^i(a_2) \leqslant \cdots \leqslant g^i(a_L)$. Similarly, for each $s \geqslant 1$, $e(s) = \frac{\lambda}{1-(1-\lambda)^L} \sum_{t=1}^{L} (1-\lambda)^{t-1} g(a_t(s))$, with $a_1(s), \ldots, a_L(s) \in A$.

We define a path $(b_t)_{t \geqslant 1}$ of pure actions in A by consecutive blocks of length L: in each of the first K blocks, play $a_1, \ldots, a_L$, then at each block $k \geqslant K+1$ play $a_1(k-K), \ldots, a_L(k-K)$. Notice that $1 - \mu = (1-\lambda)^L$ and $\sum_{t=1}^{KL} \lambda(1-\lambda)^{t-1} = (1-(1-\mu)^K)$. By construction we have

$$d = (1 - (1-\mu)^K)c + (1-\mu)^K \sum_{s=1}^{\infty} \mu(1-\mu)^{s-1} e(s)$$

$$= \sum_{t=1}^{\infty} \lambda(1-\lambda)^{t-1} g(b_t).$$

For $T \geqslant 1$, denote by $d(T)$ the continuation payoff $\sum_{t=T}^{\infty} \lambda(1-\lambda)^{t-T} g(b_t)$. If $T = kL+1$ for some $k \geqslant 0$, $d(T)$ is a convex combination of $c, e(1), \ldots, e(s), \ldots$, so $\|d - d(T)\| \leqslant 3(N+3)\varepsilon$, and if $k \geqslant K$ we have $d^i(T) \geqslant d^i + \varepsilon$. Moreover, since $c^i \leqslant d^i$, we have $d^i \leqslant d^i(L+1) \leqslant d^i(2L+1) \leqslant \cdots \leqslant d^i(KL+1)$. Assume that λ is also small enough so that for every sequence $(m_t)_{t \geqslant 1}$ in A and $l = 1, \ldots, L$, the norm between $\sum_{t=1}^{\infty} \lambda(1-\lambda)^{t-1} g(m_t)$ and the continuation payoff $\sum_{t=l}^{\infty} \lambda(1-\lambda)^{t-l} g(m_t)$ is at most ε. So for any $T \geqslant 1$, $\|d - d(T)\| \leqslant (3N+10)\varepsilon$. If $T \geqslant KL+1$, $d^i(T) \geqslant d^i + \varepsilon - \varepsilon = d^i$. Assume finally that $T \in [(k-1)L+1, kL]$ for some $k = 1, \ldots, K$. Since $g^i(a_l)_{l=1,\ldots,L}$ is non-decreasing and $d^i(kL+1) \geqslant c^i$, we have $d^i(T) \geqslant d^i((k-1)L+1)$, and this is at least d^i.

Exercise 12

Let us represent a 2-player quitting game as follows: if player 1 quits alone, the payoff $r(\{1\})$ is (a, b), if player 2 quits alone the payoff is (c, d) and if both players quit for the first time at the same stage, the payoff is (e, f).

	C_2	Q_2
C_1	$(0,0)_\circlearrowleft$	$(c,d)_*$
Q_1	$(a,b)_*$	$(e,f)_*$

If $a \leqslant 0$ and $d \leqslant 0$, never quitting is a Nash equilibrium, so we assume without loss of generality that $a = 1$. We also assume w.l.o.g. that $b < f$, $c > e$, and $d < 0$ (otherwise it is easy to construct a pure Nash equilibrium with payoff $(1, b)$, (e, f) or (c, d)). Now, either

• $b \geqslant d$: in this case the strategy profile where player 1 quits independently with probability $\varepsilon > 0$ at each stage, and player 2 always continues, is an $\varepsilon(f - d)$-Nash equilibrium of the quitting game, or

• $0 > d > b$: the strategy profile where player 1 quits independently with probability ε at each stage, and player 2 quits at every stage, is an $\varepsilon(c - e)$-Nash equilibrium of the quitting game (for $\varepsilon > 0$ small enough).

References

1. Abreu, D., Pearce, D., Stacchetti, E.: Toward a theory of discounted repeated games with imperfect monitoring. Econometrica **58**, 1041–1063 (1990)
2. Aumann, R.J.: Mixed and behaviour strategies in infinite extensive games. In: Dresher, M., Shapley, L.S., Tucker, A.W. (eds.) Advances in Game Theory. Annals of Mathematics Study, vol. 52, pp. 627–650. Princeton University Press, Princeton (1964)
3. Aumann, R.J.: Subjectivity and correlation in randomized strategies. J. Math. Econ. **1**, 67–96 (1974)
4. Aumann, R.J.: Survey of repeated games. Essays in Game Theory and Mathematical Economics in Honor of Oskar Morgenstern, pp. 11–42. Bibliographisches Institut, Mannheim (1981)
5. Aumann, R.J.: Nash equilibria are not self-enforcing. In: Gabszewicz, J.-J., Richard, J.-F., Wolsey, L. (eds.) Economic Decision Making: Games, Econometrics and Optimisation (Essays in Honor of Jacques Dreze), pp. 201–206. Elsevier Science Publishers, Amsterdam (1990)
6. Aumann, R.J.: Backward induction and common knowledge of rationality. Games Econ. Behav. **8**, 6–19 (1995)
7. Aumann, R.J.: On the centipede game. Games Econ. Behav. **23**, 97–105 (1998)
8. Aumann, R.J., Hart, S.: Bi-convexity and bi-martingales. Isr. J. Math. **54**, 159–180 (1986)
9. Aumann, R.J., Hart, S. (eds.): Handbook of Game Theory, 1. North-Holland, Amsterdam (1992)
10. Aumann, R.J., Hart, S. (eds.): Handbook of Game Theory, 2. North-Holland, Amsterdam (1994)
11. Aumann, R.J., Hart, S. (eds.): Handbook of Game Theory, 3. North-Holland, Amsterdam (2002)
12. Aumann, R.J., Maschler, M.: Some thoughts on the minmax principle. Manag. Sci. **18**, 53–63 (1972)
13. Aumann, R.J., Maschler, M.: Repeated Games with Incomplete Information, p. 1995. M.I.T. Press, Cambridge (1995)
14. Aumann, R.J., Shapley, L.S.: Long-term competition-a game theoretic analysis. In: Megiddo, N. (ed.) Essays on Game Theory, pp. 1–15. Springer, Berlin (1994)
15. Aumann, R.J., Sorin, S.: Cooperation and bounded recall. Games Econ. Behav. **1**, 5–39 (1989)
16. Barron, E.N.: Game Theory, an Introduction. Wiley, New York (2008)
17. Başar, T., Olsder, G.J.: Dynamic Noncooperative Game Theory. Classics in Applied Mathematics, vol. 23. SIAM, Philadelphia (1999)

© Springer Nature Switzerland AG 2019
R. Laraki et al., *Mathematical Foundations of Game Theory*, Universitext,
https://doi.org/10.1007/978-3-030-26646-2

18. Benedetti, R., Risler, J.-J.: Real Algebraic and Semi-algebraic Sets. Hermann, Paris (1990)
19. Benoit, J.-P., Krishna, V.: Finitely repeated games. Econometrica **53**, 905–922 (1985)
20. Benoit, J.-P., Krishna, V.: Nash equilibria of finitely repeated games. Int. J. Game Theory **16**, 197–204 (1987)
21. Ben-Porath, E., Dekel, E.: Signaling future actions and the potential for sacrifice. J. Econ. Theory **57**, 36–51 (1992)
22. Berge, C.: Espaces Topologiques. Fonctions Multivoques. Dunod, Paris (1966). English edition: Berge, C. Topological Spaces: Including a Treatment of Multi-valued Functions, Vector Spaces and Convexity, Macmillan (1963)
23. Bernheim, D.: Rationalizable strategic behavior. Econometrica **52**, 1007–1028 (1984)
24. Bich, P., Laraki, R.: On the existence of approximate equilibria and sharing rule solutions in discontinuous games. Theor. Econ. **12**, 79–108 (2017)
25. Billingsley, P.: Convergence of Probability Measures. Wiley, New York (1999)
26. Bishop, D.T., Cannings, C.: A generalized war of attrition. J. Theor. Biol. **70**, 85–124 (1978)
27. Blackwell, D.: An analog of the minmax theorem for vector payoffs. Pac. J. Math. **6**, 1–8 (1956)
28. Blackwell, D., Ferguson, T.: The big match. Ann. Math. Stat. **33**, 882–886 (1968)
29. Bolte, J., Gaubert, S., Vigeral, G.: Definable zero-sum stochastic games. Math. Oper. Res. **40**, 171–191 (2015)
30. Border, K.C.: Fixed Point Theorems with Applications to Economics and Game Theory. Cambridge University Press, Cambridge (1999)
31. Borel, E.: La théorie du jeu et les équations intégrales à noyau symétrique gauche. C.R.A.S. **173**, 1304–1308 (1921). English edition: Borel, E. On games that involve chance and the skill of the players, Econometrica, **21**, 97–110 (1951)
32. Brouwer, L.E.J.: Uber Abbildung von Mannigfaltikeiten. Math. Ann. **71**, 97–115 (1910)
33. Brown, G.W.: Iterative solutions of games by fictitious play. In: Koopmans, T.C. (ed.) Activity Analysis of Production and Allocation, pp. 374–376. Wiley, New York (1951)
34. Brown, G.W., von Neumann, J.: Solutions of games by differential equations. In: Kuhn, H.W., Tucker, A.W. (eds.) Contributions to the Theory of Games, I. Annals of Mathematical Studies, vol. 24, pp. 73–79. Princeton University Press, Princeton (1950)
35. Cesa-Bianchi, N., Lugosi, G.: Prediction. Learning and Games. Cambridge University Press, Cambridge (2006)
36. Chen, X., Deng, X.: Settling the Complexity of Two-Player Nash Equilibrium. In: FOCS, pp. 261–272 (2006)
37. Choquet, G.: Lectures on Analysis, vols. 1–3. WA Benjamin, Inc., New York (1969)
38. Cournot, A.-A.: Recherches sur les principes mathématiques de la théorie des richesses. Hachette, Paris (1938)
39. Dafermos, S.C.: Traffic equilibrium and variational inequalities. Transp. Sci. **14**, 42–54 (1980)
40. Dekel, E., Siniscalchi, M.: Epistemic game theory. In: Young, H.P., Zamir, S. (eds.) Handbook of Game Theory, vol. 4, pp. 619–702. North Holland, Amsterdam (2015)
41. De Meyer, B.: Repeated games and partial differential equations. Math. Oper. Res. **21**, 209–236 (1996)
42. De Meyer, B.: Repeated games, duality and the central limit theorem. Math. Oper. Res. **21**, 237–251 (1996)
43. Demichelis, S., Germano, F.: On the indices of zeros of Nash fields. J. Econ. Theory **92**, 192–217 (2000)
44. Demichelis, S., Germano, F.: On knots and dynamics in games. Games Econ. Behav. **41**, 46–60 (2002)
45. Demichelis, S., Ritzberger, K.: From evolutionary to strategic stability. J. Econ. Theory **113**, 51–75 (2003)
46. Dresher, M.: Games of Strategy. Prentice-Hall, London (1961)
47. Dresher, M., Shapley, L.S., Tucker, A.W. (eds.): Advances in Game Theory. Annals of Mathematical Studies, vol. 52. Princeton University Press, Princeton (1964)

48. Dresher, M., Tucker, A.W., Wolfe, P. (eds.) Contributions to the Theory of Games, III. Annals of Mathematics Studies, vol. 39. Princeton University Press, Princeton (1957)
49. Dubey, P.: Inefficiency of Nash equilibria. Math. Oper. Res. **11**, 1–8 (1986)
50. Dunford, N., Schwartz, J.T.: Linear Operators. Wiley-Interscience, New York (1988)
51. Dupuis, P., Nagurney, A.: Dynamical systems and variational inequalities. Ann. Oper. Res. **44**, 9–42 (1993)
52. Elmes, S., Reny, P.: On the strategic equivalence of extensive form games. J. Econ. Theory **62**, 1–23 (1994)
53. Fan, K.: Fixed-points and minmax theorems in locally convex topological linear spaces. Proc. Natl. Acad. Sci. USA **38**, 121–126 (1952)
54. Fan, K.: Minmax theorems. Proc. Natl. Acad. Sci. USA **39**, 42–47 (1953)
55. Flesch, J., Thuijsman, F., Vrieze, K.: Cyclic Markov equilibria in stochastic games. Int. J. Game Theory **26**, 303–314 (1997)
56. Forges, F.: An approach to communication equilibria. Econometrica **54**, 1375–1385 (1986)
57. Forges, F.: Correlated equilibrium in two person zero-sum games. Econometrica **58**, 515–516 (1990)
58. Forges, F.: Universal mechanisms. Econometrica **58**, 1341–1364 (1990)
59. Forges, F.: Repeated games of incomplete information: non-zero-sum. In: Aumann, R.J., Hart, S. (eds.) Handbook of Game Theory, 1, pp. 155–177. Elsevier Science Publishers, Amsterdam (1992)
60. Forges, F., Mertens, J.-F., Neyman, A.: A counterexample to the Folk theorem with discounting. Econ. Lett. **20**, 7 (1986)
61. Foster, D., Vohra, R.: A randomization rule for selecting forecasts. Oper. Res. **41**, 704–707 (1993)
62. Foster, D., Vohra, R.: Calibrated leaning and correlated equilibria. Games Econ. Behav. **21**, 40–55 (1997)
63. Foster, D., Vohra, R.: Asymptotic calibration. Biometrika **85**, 379–390 (1998)
64. Foster, D., Vohra, R.: Regret in the on-line decision problem. Games Econ. Behav. **29**, 7–35 (1999)
65. Fudenberg, D., Levine, D.K.: Consistency and cautious fictitious play. J. Econ. Dyn. Control **19**, 1065–1089 (1995)
66. Fudenberg, D., Levine, D.K.: The Theory of Learning in Games. MIT Press, Cambridge (1998)
67. Fudenberg, D., Levine, D.K.: Conditional universal consistency. Games Econ. Behav. **29**, 104–130 (1999)
68. Fudenberg, D., Levine, D.K., Maskin, E.: The Folk theorem with imperfect public information. Econometrica **62**, 997–1039 (1994)
69. Fudenberg, D., Maskin, E.: The Folk theorem in repeated games with discounting or with incomplete information. Econometrica **54**, 533–554 (1986)
70. Fudenberg, D., Maskin, E.: On the dispensability of public randomization in discounted repeated games. J. Econ. Theory **53**, 428–438 (1991)
71. Fudenberg, D., Tirole, J.: Perfect Bayesian equilibrium and sequential equilibrium. J. Econ. Theory **53**, 236–260 (1991)
72. Fudenberg, D., Tirole, J.: Game Theory. M.I.T. Press, Cambridge (1992)
73. Gale, D., Kuhn, H., Tucker, A.W.: On symmetric games. In: Kuhn, H.W., Tucker, A.W. (eds.) Contributions to the Theory of Games, I. Annals of Mathematical Studies, vol. 24, pp. 81–87. Princeton University Press, Princeton (1950)
74. Gale, D., Shapley, L.S.: College admissions and the stability of marriage. Am. Math. Mon. **69**, 9–15 (1962)
75. Gale, D., Stewart, F.M.: Infinite games with perfect information. In: Kuhn, H., Tucker, A.W. (eds.) Contributions to the Theory of Games, II. Annals of Mathematical Study, vol. 28, pp. 245–266. Princeton University Press, Princeton (1953)
76. Gensbittel, F., Renault, J.: The value of Markov chain games with lack of information on both sides. Math. Oper. Res. **40**, 820–841 (2015)

77. Gillette, D.: Stochastic games with zero-stop probabilities. In: Dresher, M., Tucker, A.W., Wolfe, P. (eds.) Contributions to the Theory of Games, III. Annals of Mathematical Study, vol. 39, pp. 179–187. Princeton University Press, Princeton (1957)

78. Gilboa, I., Matsui, A.: Social stability and equilibrium. Econometrica **58**, 859–867 (1991)

79. Glicksberg, I.: A further generalization of the Kakutani fixed point theorem, with applications to Nash equilibrium points. Proc. Am. Math. Soc. **3**, 170–174 (1952)

80. Gonzalez-Diaz, J., Garcia-Jurado, I., Fiestras-Janeiro, M.G.: An Introductory Course on Mathematical Game Theory. GSM, vol. 115. AMS, Providence (2010)

81. Gossner, O.: The Folk theorem for finitely repeated games with mixed strategies. Int. J. Game Theory **24**, 95–107 (1995)

82. Gossner, O., Tomala, T.: Secret correlation in repeated games with imperfect monitoring. Math. Oper. Res. **32**, 413–424 (2007)

83. Govindan, S., Wilson, R.: Equivalence and invariance of the index and degree of Nash equilibria. Games Econ. Behav. **21**, 56–61 (1997)

84. Govindan, S., Wilson, R.: Direct proofs of generic finiteness of Nash equilibrium outcomes. Econometrica **69**, 765–769 (2001)

85. Govindan, S., Wilson, R.: Essential equilibria. Proc. Natl. Acad. Sci. **102**, 15706–15711 (2005)

86. Govindan, S., Wilson, R.: Axiomatic equilibrium selection in generic two-player games in extensive form. Econometrica **80**, 1639–1699 (2012)

87. Gul, F., Pearce, D., Stachetti, E.: A bound on the proportion of pure strategy equilibria in generic games. Math. Oper. Res. **18**, 548–552 (1993)

88. Hammerstein, P., Selten, R.: Game theory and evolutionary biology. In: Aumann, R.J., Hart, S. (eds.) Handbook of Game Theory, 2, pp. 929–993. North Holland, Amsterdam (1994)

89. Hannan, J.: Approximation to Bayes risk in repeated plays. In: Dresher, M., Tucker, A.W., Wolfe, P. (eds.) Contributions to the Theory of Games, III. Annals of Mathematical Study, vol. 39, pp. 97–139. Princeton University Press, Princeton (1957)

90. Harris, C.: On the rate of convergence of continuous time fictitious play. Games Econ. Behav. **22**, 238–259 (1998)

91. Harsanyi, J.C.: Games with incomplete information played by 'Bayesian' players, parts I–III. Manag. Sci. **8**, 159–182, 320–334, 486–502 (1967–1968)

92. Harsanyi, J.C.: Games with randomly disturbed payoffs: a new rationale for mixed strategy equilibrium points. Int. J. Game Theory **2**, 1–23 (1973)

93. Harsanyi, J.C.: Oddness of the number of equilibrium points: a new proof. Int. J. Game Theory **2**, 235–250 (1973)

94. Hart, S.: Nonzero-sum two-person repeated games with incomplete information. Math. Oper. Res. **10**, 117–153 (1985)

95. Hart, S.: Adaptive heuristics. Econometrica **73**, 1401–1430 (2005)

96. Hart, S., Mas-Colell, A.: Simple Adaptive Strategies: from Regret-Matching to Uncoupled Dynamics. World Scientific Publishing, Singapore (2013)

97. Hart, S., Schmeidler, D.: Existence of correlated equilibria. Math. Oper. Res. **14**, 18–25 (1989)

98. Hillas, J., Kohlberg, E.: Foundations of strategic equilbrium. In: Aumann, R.J., Hart, S. (eds.) Handbook of Game Theory, 3, pp. 1595–1663. North Holland, Amsterdam (2002)

99. Hofbauer, J.: From Nash and Brown to Maynard Smith: Equilibria. Dynamics and ESS. Selection **1**, 81–88 (2000)

100. Hofbauer, J., Sandholm, W.H.: Stable games and their dynamics. J. Econ. Theory **144**, 1665–1693 (2009)

101. Hofbauer, J., Sorin, S.: Best response dynamics for continuous zero-sum games. Discret. Contin. Dyn. Syst.-Ser. B **6**, 215–224 (2006)

102. Hofbauer, J., Sigmund, K.: Evolutionary Games and Population Dynamics, p. 1998. Cambridge University Press, Cambridge (1998)

103. Isbell, J.R.: Finitary games. In: Dresher, M., Tucker, A.W., Wolfe, P. (eds.) Contributions to the Theory of Games, III. Annals of Mathematical Study, vol. 39, pp. 79–96. Princeton University Press, Princeton (1957)

104. Kakutani, S.: A generalization of Brouwer's fixed point theorem. Duke Math. J. **8**, 416–427 (1941)
105. Kamien, M., Tauman, Y., Zamir, S.: On the value of information in a strategic conflict. Games Econ. Behav. **2**, 129–153 (1990)
106. Karlin, A., Peres, Y.: Game Theory. Alive. AMS, Providence (2017)
107. Kelley, J.L., Namioka, I.: Linear Topological Spaces. Van Nostrand, Princeton (1963)
108. Kohlberg, E., Mertens, J.-F.: On the strategic stability of equilibria. Econometrica **54**, 1003–1037 (1986)
109. Koutsoupias, E., Papadimitriou, C.: Worst-case equilibria. Comput. Sci. Rev. **3**, 65–69 (2009)
110. Kreps, D., Sobel, J.: Signalling. In: Aumann, R.J., Hart, S. (eds.) Handbook of Game Theory, 2, pp. 849–867. North Holland, Amsterdam (1998)
111. Kreps, D., Wilson, R.: Sequential equilibria. Econometrica **50**, 863–894 (1982)
112. Kuhn, H.W.: Extensive games and the problem of information. In: Kuhn, H.W., Tucker, A.W. (eds.) Contributions to the Theory of Games, II. Annals of Mathematical Studies, vol. 28, pp. 193–216. Princeton University Press, Princeton (1953)
113. Kuhn, H.W., Tucker, A.W. (eds.): Contributions to the Theory of Games, I. Annals of Mathematics Studies, vol. 24. Princeton University Press, Princeton (1950)
114. Kuhn, H.W., Tucker, A.W. (eds.): Contributions to the Theory of Games, II. Annals of Mathematics Studies, vol. 28. Princeton University Press, Princeton (1953)
115. Laraki, R.: Variational inequalities, system of functional equations and incomplete information repeated games. SIAM J. Control Optim. **40**, 516–524 (2001)
116. Laraki, R.: The splitting game and applications. Int. J. Game Theory **30**, 359–376 (2001)
117. Laraki, R., Renault, J., Tomala, T.: Théorie des Jeux, X-UPS 2006, Editions de l'Ecole Polytechnique (2007)
118. Laraki, R., Sorin, S.: Advances in zero-sum dynamic games. In: Young, P., Zamir, S. (eds.) Handbook of Game Theory, 4, pp. 27–93. Elsevier, Amsterdam (2015)
119. Lehrer, E.: Nash equilibria of n player repeated games with semi-standard information. Int. J. Game Theory **19**, 191–217 (1989)
120. Lehrer, E.: On the equilibrium payoffs set of two-player repeated games with imperfect monitoring. Int. J. Game Theory **20**, 211–226 (1992)
121. Lemke, C.E., Howson, J.T.: Equilibrium points of bimatrix games. SIAM J. **12**, 413–423 (1964)
122. Loomis, L.H.: On a theorem of von Neumann. Proc. Natl. Acad. Sci. USA **32**, 213–215 (1946)
123. Maschler, M., Solan, E., Zamir, S.: Game Theory. Cambridge University Press, Cambridge (2013)
124. Mailath, G.J., Samuelson, L.: Repeated Games and Reputations: Long-Run Relationships. Oxford University Press, Oxford (2006)
125. Maitra, A., Sudderth, W.: Discrete Gambling and Stochastic Games. Springer, Berlin (1996)
126. Martin, D.A.: Borel determinacy. Ann. Math. **102**, 363–371 (1975)
127. Martin, D.A.: A purely inductive proof of Borel determinacy. Recursion Theory, Proceedings of Symposia in Pure Mathematics **42**, 303–308 (1985)
128. Mas-Colell, A.: On a theorem of Schmeidler. J. Math. Econ. **3**, 201–206 (1984)
129. Mas-Colell, A., Whinston, M., Green, J.: Microeconomic Theory. Oxford University Press, Oxford (1995)
130. Maynard Smith, J.: Evolution and the Theory of Games. Cambridge University Press, Cambridge (1981)
131. McLennan, A., Tourky, R.: From imitation games to Kakutani, preprint (2006)
132. Mertens, J.-F.: Repeated Games. In: Proceedings of the International Congress of Mathematicians (Berkeley), 1986, pp. 1528–1577. American Mathematical Society, Providence (1987)
133. Mertens, J.-F.: Stable equilibria - a reformulation, part I. Math. Oper. Res. **14**, 575–624 (1989)
134. Mertens, J.-F.: Stable equilibria - a reformulation, part II. Math. Oper. Res. **16**, 694–753 (1991)
135. Mertens, J.-F.: Stochastic games. In: Aumann, R.J., Hart, S. (eds.) Handbook of Game Theory, 3, pp. 1809–1832. North Holland, Amsterdam (2002)
136. Mertens, J.-F., Neyman, A.: Stochastic games. Int. J. Game Theory **10**, 53–66 (1981)

137. Mertens, J.-F., Sorin, S., Zamir, S.: Repeated Games. Cambridge University Press, Cambridge (2015)
138. Mertens, J.-F., Zamir, S.: The value of two-person zero-sum repeated games with lack of information on both sides. Int. J. Game Theory **1**, 39–64 (1971)
139. Mertens, J.-F., Zamir, S.: The normal distribution and repeated games. Int. J. Game Theory **5**, 187–197 (1976)
140. Mertens, J.-F., Zamir, S.: A duality theorem on a pair of simultaneous functional equations. J. Math. Anal. Appl. **60**, 550–558 (1977)
141. Milgrom, P., Weber, R.: Distributional strategies for games with incomplete information. Math. Oper. Res. **10**, 619–632 (1985)
142. Mills, H.D.: Marginal value of matrix games and linear programs. In: Kuhn, H.W., Tucker, A.W. (eds.) Linear Inequalities and Related Systems. Annals of Mathematical Studies, vol. 38, pp. 183–193. Princeton University Press, Princeton (1956)
143. Milnor, J.W.: Topology from the Differentiable Viewpoint. Princeton University Press, Princeton (1965)
144. Monderer, D., Samet, D., Sela, A.: Belief affirming in learning processes. J. Econ. Theory **73**, 438–452 (1997)
145. Monderer, D., Sela, A.: A 2×2 game without the fictitious play. Games Econ. Behav. **14**, 144–148 (1996)
146. Monderer, D., Shapley, L.S.: Potential games. Games Econ. Behav. **14**, 124–143 (1996)
147. de Montmort, P.R.: Essay d'analyse sur les jeux de hazard, 2nd edn. chez Jacque Quillau, Paris (1713)
148. Myerson, R.: Refinements of the Nash equilibrium concept. Int. J. Game Theory **7**, 73–80 (1978)
149. Myerson, R.: Game Theory. Harvard University Press, Harvard (1991)
150. Myerson, R.: Communication, correlated equilibria and incentive compatibility. In: Aumann, R.J., Hart, S. (eds.) Handbook of Game Theory, 2, pp. 827–847. North Holland, Amsterdam (1998)
151. Nash, J.: Equilibrium points in n-person games. Proc. Natl. Acad. Sci. **36**, 48–49 (1950)
152. Nash, J.: The bargaining problem. Econometrica **18**, 155–162 (1950)
153. Nash, J.: Non-cooperative games. Ann. Math. **54**, 286–295 (1951)
154. Neyman, A., Sorin, S.: Stochastic Games and Applications. NATO Science Series. Kluwer Academic Publishers, Dordrecht (2003)
155. Nikaido, H., Isoda, K.: Note on non-cooperative convex games. Pac. J. Math. **5**, 807–815 (1955)
156. Nisan, N., Roughgarden, T., Tardos, E., Vazirani, V.: Algorithmic Game Theory. Cambridge University Press, Cambridge (2007)
157. Osborne, M.J., Rubinstein, A.: A Course in Game Theory. MIT Press, Cambridge (1994)
158. Owen, G.: Game Theory, 3rd edn. Academic, New York (1995)
159. Parthasarathy, K.R.: Probability Measures on Metric Spaces. Academic, New York (1967)
160. Parthasarathy, T., Raghavan, T.E.S.: Some Topics in Two-Person Games. American Elsevier, Amsterdam (1971)
161. Pearce, D.: Rationalizable strategic behavior and the problem of perfection. Econometrica **52**, 1029–50 (1984)
162. Peleg, B.: Equilibrium points for games with infinitely many players. J. Lond. Math. Soc. **44**, 292–294 (1969)
163. Prokopovych, P.: On equilibrium existence in payoff secure games. Econ. Theory **48**, 5–16 (2011)
164. Renault, J.: 2-player repeated games with lack of information on one side and state independent signalling. Math. Oper. Res. **25**, 552–572 (2000)
165. Renault, J.: Repeated games with incomplete information. In: Meyers, R.A. (ed.) Encyclopedia of Complexity and System Science (2009). ISBN: 978-0-387-75888-6
166. Renault, J.: Uniform value in dynamic programming. J. Eur. Math. Soc. **13**, 309–330 (2011)

167. Renault, J.: The value of repeated games with an informed controller. Math. Oper. Res. **37**, 154–179 (2012)
168. Renault, J.: Zero-sum stochastic games. Tutorial for Stochastic Methods in Game Theory, Institute of Mathematical Sciences, National university of Singapore (2015)
169. Renault, J., Tomala, T.: Repeated proximity games. Int. J. Game Theory **27**, 539–559 (1998)
170. Renault, J., Tomala, T.: Communication equilibrium payoffs of repeated games with imperfect monitoring. Games Econ. Behav. **49**, 313–344 (2004)
171. Renault, J., Venel, X.: A distance for probability spaces, and long-term values in Markov decision processes and repeated games. Math. Oper. Res. **42**, 349–376 (2017)
172. Renault, J., Ziliotto, B.: Hidden stochastic games and limit equilibrium payoffs (2014). arXiv:1407.3028
173. Reny, P.: Rational behaviour in extensive-form games. Can. J. Econ. **28**, 1–16 (1995)
174. Reny, P.: On the existence of pure and mixed Nash equilibria in discontinuous games. Econometrica **67**, 1029–1056 (1999)
175. Ritzberger, K.: The theory of normal form games from the differentiable viewpoint. Int. J. Game Theory **23**, 207–236 (1994)
176. Ritzberger, K.: Foundations of Non-cooperative Game Theory. Oxford University Press, Oxford (2002)
177. Robinson, J.: An iterative method of solving a game. Ann. Math. **54**, 296–301 (1951)
178. Rosen, J.B.: Existence and uniqueness of equilibrium points for concave n-person games. Econometrica **33**, 520–534 (1965)
179. Rosenberg, D., Solan, E., Vieille, N.: Blackwell optimality in Markov decision processes with partial observation. Ann. Stat. **30**, 1178–1193 (2002)
180. Rosenberg, D., Sorin, S.: An operator approach to zero-sum repeated games. Isr. J. Math. **121**, 221–246 (2001)
181. Rosenthal, R.: Games of perfect information, predatory pricing and the chain store paradox. J. Econ. Theory **25**, 92–100 (1982)
182. Rubinstein, A.: Perfect equilibrium in a bargaining model. Econometrica **50**, 97–110 (1982)
183. Rubinstein, A.: Equilibrium in supergames. In: Meggiddo, N. (ed.) Essay in Game Theory in Honor of M. Maschler, pp. 17–28. Springer, Berlin (1994). (original paper: Equilibrium in supergames, Center for Research in Mathematical Economics and Game Theory, Research Memorandum 25.)
184. Sandholm, W.H.: Potential games with continuous player sets. J. Econ. Theory **97**, 81–108 (2001)
185. Sandholm, W.H.: Population Games and Evolutionary Dynamics. MIT Press, Cambridge (2011)
186. Schmeidler, D.: Equilibrium points of nonatomic games. J. Stat. Phys. **7**, 295–300 (1973)
187. Selten, R.: Re-examination of the perfectness concept for equilibrium points in extensive games. Int. J. Game Theory **4**, 25–55 (1975)
188. Shapley, L.S.: Stochastic games. Proc. Natl. Acad. Sci. USA **39**, 1095–1100 (1953)
189. Shapley, L.S.: Some topics in two-person games. In: Dresher, M., Shapley, L.S., Tucker, A.W. (eds.) Advances in Game Theory. Annals of Mathematical Studies, vol. 52, pp. 1–28. Princeton University Press, Princeton (1964)
190. Simon, R.S., Spież, S., Toruńczyk, H.: The existence of equilibria in certain games, separation for families of convex functions and a theorem of Borsuk-Ulam type. Isr. J. Math. **92**, 1–21 (1995)
191. Sion, M.: On general minmax theorems. Pac. J. Math. **8**, 171–176 (1958)
192. Sion, M., Wolfe, P.: On a game without a value. In: Dresher, M., Tucker, A.W., Wolfe, P. (eds.) Contributions to the Theory of Games, III. Annals of Mathematical Studies, vol. 39, pp. 299–306. Princeton University Press, Princeton (1957)
193. Smith, M.J.: The stability of a dynamic model of traffic assignment - an application of a method of Lyapunov. Transp. Sci. **18**, 245–252 (1984)
194. Solan, E.: Three-player absorbing games. Math. Oper. Res. **24**, 669–698 (1999)
195. Solan, E.: Stochastic games. Encyclopedia of Database Systems. Springer, Berlin (2009)

196. Solan, E., Vieille, N.: Quitting games. Math. Oper. Res. **26**, 265–285 (2001)
197. Sorin, S.: Some results on the existence of Nash equilibria for non-zero-sum games with incomplete information. Int. J. Game Theory **12**, 193–205 (1983)
198. Sorin, S.: On repeated games with complete information. Math. Oper. Res. **11**, 147–160 (1986)
199. Sorin, S.: Asymptotic properties of a non-zero sum stochastic game. Int. J. Game Theory **15**, 101–107 (1986)
200. Sorin, S.: Information and rationality: some comments. Annales d' Economie et de Statistique **25**(26), 315–325 (1992)
201. Sorin, S.: Repeated games with complete information. In: Aumann, R.J., Hart, S. (eds.) Handbook of Game Theory, 1, pp. 71–107. North Holland, Amsterdam (1992)
202. Sorin, S.: Merging, reputation and repeated games with incomplete information. Games Econ. Behav. **29**, 274–308 (1999)
203. Sorin, S.: A First Course on Zero-Sum Repeated Games. Mathématiques et Applications, vol. 37. Springer, Berlin (2002)
204. Sorin, S., Wang, C.: Finite composite games: equilibria and dynamics. J. Dyn. Games **3**, 101–120 (2016)
205. Sorin, S., Zamir, S.: A 2-person game with lack of information on 1 and 1/2 sides. Math. Oper. Res. **10**, 17–23 (1985)
206. Sperner, E.: Neuer Beweis fur die Invarianz der Dimensionszahl und des Gebietes. Abh. Math. Sem. Univ. Hambg. **6**, 265–272 (1928)
207. Swinkels, J.M.: Adjustment dynamics and rational play in games. Games Econ. Behav. **5**, 455–484 (1993)
208. Tarski, A.: A lattice theoretical fixed point theorem and its applications. Pac. J. Math. **5**, 285–308 (1955)
209. Taylor, P.D., Jonker, L.B.: Evolutionary stable strategies and game dynamics. Math. Biosci. **40**, 145–156 (1978)
210. Thompson, F.B.: Equivalence of Games in Extensive Form. Research Memorandum, vol. 759. The Rand Corporation (1952)
211. Tomala, T.: Nash equilibria of repeated games with observable payoff vectors. Games Econ. Behav. **28**, 310–324 (1999)
212. Topkis, D.: Equilibrium points in nonzero-sum n person submodular games. SIAM J. Control Optim. **17**, 773–787 (1979)
213. Van Damme, E.: A relation between perfect equilibria in extensive form games and proper equilibria in normal form games. Int. J. Game Theory **13**, 1–13 (1984)
214. Van Damme, E.: Stability and Perfection of Nash Equilibria. Springer, Berlin (1987)
215. Van Damme, E.: Refinement of Nash equilibrium. In: Laffond, J.-J. (ed.) Advances in Economic Theory (6th Congress Econometric Society), pp. 32–75. Cambridge University Press, Cambridge (1992)
216. Van Damme, E.: Evolutionary game theory. Eur. Econ. Rev. **34**, 847–858 (1994)
217. Van Damme, E.: Strategic equilibrium. In: Aumann, R.J., Hart, S. (eds.) Handbook of Game Theory, vol. 3, pp. 1521–1596. North Holland, Amsterdam (2002)
218. Vieille, N.: Two-player stochastic games I: a reduction. Isr. J. Math. **119**, 55–91 (2000)
219. Vieille, N.: Two-player stochastic games II: the case of recursive games. Isr. J. Math. **119**, 93–126 (2000)
220. Vieille, N.: Stochastic games: recent results. In: Aumann, R.J., Hart, S. (eds.) Handbook of Game Theory, 3, pp. 1833–1850. North Holland, Amsterdam (2002)
221. Vigeral, G.: A zero-sum stochastic game with compact action sets and no asymptotic value. Dyn. Games Appl. **3**, 172–186 (2013)
222. Ville, J.: Sur la théorie générale des jeux où intervient l'habileté des joueurs. In: Borel, E. (ed.) Traité du Calcul des Probabilités et de ses Applications, Tome IV, pp. 105–113. Gauthier-Villars, Paris (1938)
223. Vives, X.: Nash equilibrium with strategic complementarities. J. Math. Econ. **19**, 305–321 (1990)
224. Von Neumann, J.: Zur Theorie der Gesellschaftsspiele. Math. Ann. **100**, 295–320 (1928)

225. Von Neumann, J., Morgenstern, O.: Theory of Games and Economic Behavior. Princeton University Press, Princeton (1944)
226. Wardrop, G.: Some theoretical aspects of road traffic research. Proc. Inst. Civ. Eng. **1**, 325–362 (1952)
227. Weibull, J.: Evolutionary Game Theory. MIT Press, Cambridge (1995)
228. Weyl, H.: Elementary proof of a minmax theorem due to von Neumann. In: Kuhn, H.W., Tucker, A.W. (eds.) Contributions to the Theory of Games, I. Annals of Mathematical Studies, vol. 24, pp. 19–25. Princeton University Press, Princeton (1950)
229. Young, P., Zamir, S. (eds.): Handbook of Game Theory IV. North Holland, Amsterdam (2015)
230. Zamir, S.: Repeated games of incomplete information: zero-sum. In: Aumann, R.J., Hart, S. (eds.) Handbook of Game Theory, 1, pp. 109–154. North Holland, Amsterdam (1992)
231. Zermelo, E.: Über eine Anwendring der Mengenlehrer auf die Theorie des Schachspiels. In: Proceedings of the Fifth International Congress of Mathematicians (Cambridge), 1912, vol. II, p. 501 (1912)
232. Ziliotto, B.: Zero-sum repeated games: counterexamples to the existence of the asymptotic value and the conjecture maxmin $= \lim v(n)$. Ann. Probab. **44**, 1107–1133 (2016)

Index

A

Action, 107
Agent normal form, 120
Alternative theorem, 17, 26
Antisymmetric games, 29
Approximate equilibrium, 60
Auction, 3, 61, 127
Aumann and Maschler's cav(u) theorem, 172
Aumann–Shapley Folk theorem, 162
Aumann's theorem, 133

B

Backward induction, 103, 105, 114
Bargaining, 3, 126
Battle of the sexes, 115, 119, 122, 130, 179
Bayesian games, 143
Bayesian perfect equilibrium, 116
Behavioral equilibrium, 113
Behavioral strategy, 109, 111, 112, 116, 125
    generated by a mixed strategy, 112
Benoît and Krishna's Folk theorem, 160
Berge intersection lemma, 33
Bertrand game, 165
Best reply correspondence, 7
Best response
    correspondence, 7
    domination, 46
    dynamics, 21, 89
Better reply security, 57, 59, 61
Big match game, 167
Blackwell and Ferguson's theorem, 168
Blackwell approachability, 24, 25, 42, 43, 136, 140
Blackwell's theorem, 136
Brouwer's theorem, 70

Brown–von Neumann–Nash dynamics, 88
Burning money game, 114

C

Canonical correlated equilibrium, 133
Canonical information structure, 133
Cav(u) theorem, 172
Centipede game, 105, 117
Chomp, 124
Coalitional game, 1
Coalitions, 1
Coarse equilibria (or Hannan set), 140
Common knowledge, 49, 66, 117
Compact game, 46, 48, 52, 53, 57, 59, 61, 72, 73
Complementarity in linear programming, 18
Completely labeled sub-simplex, 68
Completely mixed, 116, 117
Completely revealing strategy, 172, 204
Concave game, 55, 78
Condorcet paradox, 4
Congestion game, 88, 92
Continuous game, 46, 53
Convex game, 73
Correlated equilibrium, 129, 132, 134, 147, 148
Correlated strategy, 46, 47
Cournot competition, 55, 71, 72

D

Degree of an equilibrium, 83
Derived game, 38
Determined game, 101, 106, 124
Discontinuous game, 57, 59
Discounted game, 126, 155, 158, 178–180

© Springer Nature Switzerland AG 2019
R. Laraki et al., *Mathematical Foundations of Game Theory*, Universitext,
https://doi.org/10.1007/978-3-030-26646-2

Printed in the United States
By Bookmasters